THE BICENTENNIAL TRIBUTE
TO
AMERICAN MATHEMATICS

The
BICENTENNIAL TRIBUTE
to
AMERICAN MATHEMATICS
1776   1976

Dalton Tarwater, Editor

Papers presented at the Fifty-ninth Annual Meeting of the Mathematical Association of America commemorating the nation's bicentennial

Published and distributed by

The Mathematical Association of America

© 1977 by

The Mathematical Association of America (Incorporated)
Library of Congress Catalog Card Number 77-14706

ISBN 0-88385-424-4

Printed in the United States of America

Current printing (Last Digit):

10 9 8 7 6 5 4 3 2 1

PREFACE

It was decided in 1973 that the Mathematical Association of America would commemorate the American Bicentennial at the San Antonio meeting of the Association in January, 1976, by stressing the history of American mathematics.

Some speakers were invited to trace American mathematical history from Colonial Times to the present. Others were selected to address the meeting on various topics of historical interest to the broad mathematical community represented by the membership of the Association.

In addition to the major historical addresses, there were panel discussions on Two-Year College Mathematics in 1976; Mathematics in Our Culture; The teaching of Mathematics in College: A 1976 Perspective for the Future; and The Role of Applications in the Teaching of Undergraduate Mathematics.

In 1975, the thought began to emerge that the collection of addresses and panel discussions should be published by the Association. A committee was formed to undertake this publication. That committee was composed of Professors Edwin F. Beckenbach, Leonard Gillman, Judith Grabiner, and David Roselle, with the present writer as Chairman. Each of these members has played an important role in developing the present volume.

Dirk Struik's paper appeared earlier in *Men and Institutions in American Mathematics,* Texas Tech University Graduate Studies, No. 13, Lubbock (1976), and is included here, with permission of Texas Tech University, for the sake of completeness. Mark Kac's paper, "Probability Theory: Reflections on the Past and Speculations on the Future," is, unfortunately, unobtainable for inclusion in the present volume. The address of Paul Halmos, *et al.,* has already appeared in *The American Mathematical Monthly,* 83 (1976), pp. 503–516.

DALTON TARWATER

CONTENTS

Contents

MATHEMATICS IN COLONIAL AMERICA

Dirk J. Struik

The history of mathematics can be presented in different ways. One is the "skyline" route, in which one concentrates on the high points: the great mathematicians and the great discoveries. This is the way most books on the subject are organized. Another way is the development of mathematics as a social phenomenon, as an aid to physics, astronomy, and other sciences, or as the subject of education; here one can study also its influence on the general world outlook of a generation, a class, or a special group of men.

The "skyline" approach to this period does not lead us very far. The histories of mathematics do not deal with it. The searcher for a path to the skyline can find some satisfaction in Franklin's magic squares, Adrain's derivation of the normal error law, and Bowditch's discovery of the figures commonly named after Lissajous, but little more. The first time the skyline is reached is with Benjamin Peirce's *Linear Associative Algebras* (1870) or perhaps with some of the astronomical mathematics of G. W. Hill.

More profitable is the other approach. In this case, it leads mainly to mathematics of use in astronomy, surveying, and hydrography. We shall confine our attention to North America, inclusive of Mexico.

European mathematics came to the New World with Columbus in the form of computations with the decimal position system expressed in symbols not very different from the ones presently used. More advanced mathematics came in the wake of the Conquest, as can be seen in the booklet written by Juan Diez, probably one of Cortes' chaplains; this booklet of 1556 (reprinted in 1921) deals with assaying, but has some arithmetic and algebra suggestive of knowledge of Diez's contemporary, Cardan. Among the scholars in and around the University of Mexico, which opened in 1553, one finds men with knowledge of surveying and navigation equal to that of the best contemporary Europeans. Outstanding is Enrico Martinez,

1

a German, who, around 1600, excelled as an engineer, astronomer, linguist, and polymath in general; a kind of Mexican Stevin. He is remembered primarily for his labors on the drainage system of Mexico City, where there is a monument of him on the main plaza. Passing salute goes to Thomas Hariot, visitor to present North Carolina in the 1580's and author of a famous description and map of his discoveries in Virginia. He was a young man at the time; his fame as a mathematician came later, but had nothing to do with America.

When, after 1600, the Atlantic coast of North America was being colonized, there were many settlers with a university education, something often meaning no more, mathematically speaking, than some knowledge of the rule of three. Few of these men ever crossed the *pons asinorum*. (Euclid, Elements 15: The angles at the base of an isosceles triangle are equal.) This also may have been true for the teachers at the two newly founded colleges, the Puritan one at Cambridge, and the Jesuit one at Quebec. Some Jesuits probably knew more mathematics, or at any rate did appreciate it. In 1665 French born Martin Boutet, Sieur de St. Martin, was appointed *professor matheseos* at the Quebec college. Here was taught a course in hydrography, including the mathematics deemed necessary for navigation, surveying, and cartography. The Jesuits were a little ahead of the Puritans in mathematics, but the Puritans were more receptive to new theories, willing to listen to Copernicus and Descartes and later to Locke, whereas Quebec preferred Aristotle. Both parties paid attention to formal logic, Quebec again adhering to Aristotle, Harvard more to Ramus (the Protestant educator of Paris); but at that time logic was not thought of as a mathematical discipline.

Although I know of no published mathematical work of his, Carlos de Sigüenza y Gongara (1645-1700) of Mexico, the poet-cartographer-astronomer-historian-polemicist, should be mentioned.

The first at Harvard to show a deeper interest in mathematics was Thomas Brattle (1658-1713), merchant-astronomer, working "here alone by myself, without a meet help in respect to my studies," as he wrote to Flamsteed in 1703. He had been able to cross the asses' bridge, and that done, he wrote that the rest of geometry came easy, and trigonometry followed. He used the telescope, a gift to Harvard in 1672 by John Winthrop II (son of the first governor of Massachusetts and later himself governor of Connecticut); his observations on the famous comet of 1680 were appreciated by Newton, and he may well have been the first to determine astronomically not only the latitude (already roughly known), but also the longitude of Boston by observing a lunar eclipse. In the same period, the position of Quebec was established astronomically in 1685 by the visiting French cartographer Jean Deshaies. The correct position of Mexico City was known at least from the middle of the seventeenth century, but it was not published; thus, until late into the eighteenth century, Mexico City was placed west of

Acapulco in the Pacific. The Spanish were not eager to inform other nations about their empire.

During the intellectual stagnation of the Spanish empire and the later revival stimulated by Carlos III, there was little interest in mathematics in eighteenth century Mexico. Quebec, and especially Harvard, however, were improving their appreciation of mathematics; the leading figures were Isaac Greenwood and John Winthrop IV. Father J. P. DeBonnécamps taught hydrography at Quebec between 1743 and 1758. There was also an intellectual group at Philadelphia, its Quaker atmosphere tolerant to new ideas; here, during the first half of the century, was the impressive James Logan, magistrate, botanist, physicist, author, and aristocrat; his correspondence showed critical understanding of Huygens' dioptrics and of Newton's algebra and fluxions. A Pennsylvanian Maecenas, he encouraged the young mechanic Thomas Godfrey (of Godfrey's quadrant, predecessor of the sextant) and Benjamin Franklin. As a result of talking to Logan, Franklin set up his magic squares, first the 8 by 8, then the 16 by 16 one, although he thought them *difficiles nugae* (difficult trifles). Most people would not disagree with Franklin on this.

Greenwood was the first Hollis professor of mathematics and natural philosophy (1728), a combination that lasted until the nineteenth century. He was the first teacher in America of Newtonian philosophy, giving lectures with demonstrations of the "discoveries of the incomparable Sir Isaac Newton"; these lectures may well have included some algebra and fluxions. He wrote *Arithmetic, Vulgar and Decimal* (1729), the first separate treatise on arithmetic written by a native British-American. (The first still existing American arithmetic in English is part 2 of *The Young Man's Companion*, published as a second edition in 1710 by William and Andrew Bradford, New York. The first edition was in 1705.) He lost his Harvard job in 1737 for "gross intemperance," and spent his later years as a traveling lecturer.

John Winthrop IV, Greenwood's successor to the Hollis chair, was, luckily for Harvard and Newton's prestige, full of the social graces, and continued to teach in the Newtonian tradition until his death in 1779. His was primarily applied mathematics, especially astronomy, but he also taught pure mathematics, including fluxions. From a letter he wrote in 1764, it is known that, apart from hydrostatics, mechanics, optice, astronomy, he taught: "the elements of Geometry, together with the doctrine of Proportion, the principles of Algebra, Conic Sections, Plane and Spherical Trigonometry, with general principles of Mensuration of Planes and Solids, the use of globes, the calculations of the motions and phenomena of the heavenly bodies according to the different hypotheses of Ptolemy, Tycho Brahe and Copernicus ..." as well as cartography, surveying, and navigation. Specialization was not an eighteenth century weakness. Winthrop's library survives and reflects a good acquaintance with the mathematics of his time, except the continental mathematics of Leibniz and Euler. They had to wait at Harvard until Farrar introduced them in the early nineteenth century. Winthrop was still a typical British scholar. He also was, according to Count Rumford, who as a young man listened to some of his lectures, "an excellent and happy teacher." But his influence on the development of mathematics in America seems to have been minimal.

Another astronomer with mathematical interest was the self-taught clock-maker David Rittenhouse of Philadelphia, also an able surveyor. In later life he published some mathematical papers; the most interesting one (1793) deals with what he called "the sums of the several powers of the sines," that is,

$$\int_0^{\pi/2} \sin{}^n\psi \, d\psi,$$

all in Newton's style of writing. This was new to Rittenhouse, and Newton apparently never published these sums, but the result is at least as old as Pascal. In another paper Rittenhouse (1799) used series for a solution to Kepler's equation $M = E - e \sin E$.

DeBonnécamps' mathematics also was applied. At Quebec College he taught hydrography and made astronomical observations with instruments less satisfactory than those available to Winthrop and Rittenhouse. One of his pupils was Michel Chartier de Lotbinière, known as the architect of Fort Carillon, now Ticonderoga, built in the mathematical tradition of Vauban. Another fortress built in this school was Louisbourg.

This brings one to the military engineers and that other group of men applying mathematics in their trade: the scientifically trained surveyors and cartographers in French and British service. Well known in their days were Joseph Frédéric Wallet des Barres, Samuel Holland, Bernard Romans, and William Gerard De Brahm, all active in the last decades of the eighteenth century, the first two and especially des Barres, author of that famous cartographic work, *The Atlantic Neptune* (1780). Some of these men came from Europe and stayed, others returned after some years.[1] Joseph Bernard Chabert took accurate observations with the then modern method of lunar distances and eclipses of the Jupiter satellites along the North Atlantic Coast during 1750-51, with Louisbourg as base. He later returned during the American Revolution and, with the geodesic surveyor Jean-Charles Borda, tested chronometers. From England came the most famous

---

1. An aide-de-camp to Montcalm in the Canadian campaign of 1758–59 was young Louis-Antoine de Bougainville, a protégé of D'Alembert and already the author of a two-volume text on the integral calculus, at that time the best exposition of the continental approach. He returned to France and became famous as an explorer. Although he had no influence on American mathematics, he probably was the first one to set foot on North American soil with a thorough knowledge of continental mathematics. Think of him when you see bougainvilleas.

surveyors of America's eighteenth century, Charles Mason and Jeremiah Dixon; they were active in the determination of the Pennsylvania-Delaware-Maryland line. The influence of these men on the growth of interest in applied mathematics is difficult to estimate; it has been studied only in special cases, as in that of Mason and Dixon, and Pennsylvanian surveyors. The same holds for the foreign observers of the transit of Venus in 1769, although it is known that the participation of the Abbé Jean-Baptiste Chappe in the California expedition had some influence on Mexican science.

The new republic had several colleges (Harvard, Yale, William and Mary, Pennsylvania, Princeton, Rutgers, Bowdoin) and many academies. The college in Quebec had disappeared with the Conquest and the expulsion of the Jesuits; the latter event also was harmful to the instruction of science in the Spanish Empire. (However, only the mathematics of the young United States will concern us here.) The colleges in the U.S.A. showed little interest in mathematics. Harvard, in 1803, required for entrance the mere rudiments of arithmetic; in 1816, the whole of elementary arithmetic; and in 1819, a light knowledge of algebra. Not until 1837 was arithmetic dropped from the freshman course. The situation in other colleges was not much better. However, after 1800 some good teachers appeared and gallantly engaged to raise the level of mathematical knowledge. Irish born Robert Adrain (1775–1843), who taught at Columbia, Philadelphia, Princeton, and Rutgers; New England born John Farrar (1779–1856), who modernized mathematical instruction at Harvard; Theodore Strong, who taught at Rutgers, and some French teachers such as Claude Crozet at West Point represented a new element in American mathematics: the influence of France and its Revolution.

All through colonial days, the only European influence had been that of Great Britain with its strict Newtonian tradition. With the American Revolution came the admiration of, or at any rate the interest in, France and its advanced mathematical schools. When, under Jefferson's influence West Point was established, the Paris École Polytechnique, with its emphasis on mathematics, served as example. Strong, Farrar, and Crozet brought French mathematical texts to the attention of their students. It was Farrar, since 1807 the occupant of the Hollis chair at Harvard, who introduced into his instruction French material through English versions or translations. Between 1818 and 1829, he introduced presentations of material by Lacroix, Legendre, Bézout, Biot, and Euler, beginning with Lacroix's *Elements of Algebra.* At West Point, Crozet taught descriptive geometry, Monge's brain child.

Legendre's *Éléments de géometrie,* first published in 1794 as a then modern approach to Euclid, was one of the most influential textbooks of the period. It was translated several times into English, first by Farrar in 1819; another translation, brought out anonymously by no other than Thomas Carlyle during his early years (*ca.* 1820), was very successful in a revision of 1828 by the West Point professor Charles Davies (1798-1876). Davies wrote many other textbooks, among them *Analytical Geometry*, an English version of a book written by Pierre Bourdon. In 1843-44 Harvard first made geometry a requirement for admission, this through the influence of Benjamin Peirce (1809-1880), professor since 1833 and a prolific writer of textbooks, beginning in 1835 with a *Treatise on Plane Trigonometry*.

But it was Laplace, in particular through his *Mécanique Céleste* (5 vols, 1799 to 1825), who stimulated the awakening creativity of American mathematicians, as shown in the works of Adrain and Nathaniel Bowditch (1773–1838). Adrain's best known paper, with the derivation of the normal law of errors (1808), was inspired by Laplace, and so were his articles on the shape of the earth. Bowditch, a Salem merchant-skipper and after 1823 a well-to-do Boston insurance executive, showed his dedication to Laplace by commenting on and translating the first four volumes of *Mécanique Céleste* into English (1829 to 1839), a labor of love paid for from his own pocket.

Among Bowditch's original work can be mentioned a paper suggested by the apparent motion of the earth as seen from the moon. Here he found the figures now known as those of Lissajous. It was published in 1815 in the "Memoirs" of the *American Academy of Arts and Sciences* founded in 1780 in Boston. This shows that the time had come when mathematical papers could be published in an American periodical. The first such periodical was the *Transactions of the American Philosophical Society*, established at Philadelphia and first published in 1771. Yet all through the first half of the nineteenth century and even later there were few such periodicals. Several attempts to publish such journals, even on a modest scale, had little success, from Adrain's *The Analyst* of 1808 (with his paper on errors) to *The Cambridge Miscellany of Mathematics, Physics and Astronomy*, started in 1842 by the Harvard men Benjamin Peirce and Joseph Lovering. The time had not yet come for a deeper interest in mathematics in the many academies and the growing number of colleges.

A few other names of some importance in these first decades of the nineteenth century should be mentioned. Benjamin Banneker (1731-1806), a Maryland astronomer, and a friend of the Ellicott family of merchants and surveyors, was the first black man in America to achieve distinction in science. Charles Gill, a Yorkshire man who came to America in 1830 at the age of 25, was a teacher and an actuary; he edited a periodical *Mathematical Miscellany* (Flushing, New York, 1836-39) and made several contributions to number theory. Of great importance for the U.S. Coast Survey after 1816 was Swiss-born Frederic Hassler, and for the mathematics instruction at the Naval Academy after 1845, William Chauvenet. Other names can be found in *A History of Mathematics in America Before 1900* by Smith and Ginsburg (1944). Significantly, their chapter on the introduction of modern mathematics into the United States deals with the period of 1875 to 1900, long after our period has come to an end. Our period deals with the British influence, typical of the colonial era, and the French, typical of the mercantile and early industrial period of the Republic. The period 1875 to 1900 is that of the German influence, and of developed industrialism after the Civil War.

Bibliography

J. F. W. Des Barres, The Atlantic Neptune, published for the use of the Royal Navy of Great Britain, London, 1780.

S. A. Bedini, The Life of Benjamin Banneker, Scribners, New York, 1972, xvii + 434 pp.

J. Diez, The Sumario Compendioso of Brother Juan Diez, ed. by D. E. Smith, Ginn, Boston, 1921, 65 pp.

C. C. Gillispie, ed., Dictionary of Scientific Biography. Articles on Adrain, Bowditch, Chappe, Greenwood, Hassler, Hill; Scribners, New York, 1970.

E. De Gortari, La ciencia en la historia de México. Fondo de Cultura Económica, México, 1963, 461 pp.

I. Greenwood, Arithmetic, vulgar and decimal: with the application thereof to a variety of cases in trade and commerce, Boston, 1729.

B. Hindle, The Pursuit of Science in Revolutionary America, 1735-1789, University of North Carolina Press, Chapel Hill, 1956, 410 pp.

____, David Rittenhouse, Princeton University Press, Princeton, 1964, 394 pp.

E. R. Hogan, The beginnings of mathematics in a howling wilderness, Historia Mathematica, 1 (1974) 151-166.

____, George Baron and the Mathematical Correspondent, ibid., 3 (1976) 403-415.

____, Robert Adrain: American mathematician, ibid., 4 (1977) 157-172.

L. C. Karpinski, The History of Arithmetic, Russell and Russell, New York, 1925, 200 pp.

R. N. Lokken, ed., The scientific papers of James Logan, Trans. Amer. Philos. Soc., new series, (6) 62 (1972) 5-94.

L. W. McKeehan, Yale Science—The First Hundred Years 1701-1801, H. Schuman, New York, 1947, ix + 82 pp.

S. E. Morison, Harvard College in the Seventeenth Century, 2 vols., Harvard University Press, Cambridge, 1936a, 707 pp.

____, Three Centuries of Harvard, Harvard University Press, Cambridge, 1936b, viii + 512 pp.

B. Peirce, Linear associative algebra, Amer. J. Math., 4 (1882) 97-229. (This is the formal publication of essays published informally in 1870.)

D. Rittenhouse, Relative to a method of finding the sum of the several powers of the sines, Trans. Amer. Philos. Soc., 3 (1793) 155-156.

____, To determine the true place of a planet in an elliptical orbit, directly from the mean anomaly, by converging series, Trans. Amer. Philos. Soc., 4 (1799) 21-26.

W. H. Roever, William Chauvenet, Washington University Studies, 12 (1925) 97-117.

L. G. Simons, Introduction of algebra into American schools in the eighteenth century, Ph.D. dissertation, Columbia University, Government Printing Office, Washington, D.C., 1925, vi + 80 pp.

____, The influence of French mathematicians at the end of the eighteenth century upon the teaching of mathematics in American colleges, Isis, 15 (1931) 104-123.

D. E. Smith, Thomas Jefferson and mathematics, Scripta Math., 1 (1932) 3-14.

D. E. Smith and J. Ginsburg, A History of Mathematics in America before 1900, Mathematical Association of America with the cooperation of Open Court Publishing Co., Chicago, 1944, x + 209 pp.

D. J. Struik, Yankee Science in the Making, Brown, Boston, 1948, xiii + 430 pp.

____, Mathematicians at Ticonderoga, Scientific Monthly, 82 (1956) 236-240.

MATHEMATICS IN AMERICA: THE FIRST HUNDRED YEARS

Judith V. Grabiner

1. Introduction. The hundred years after 1776 were good years for mathematics. In France, there were Lagrange, Laplace, Cauchy; in Great Britain, Cayley, Hamilton, Sylvester; in Germany, Gauss, Riemann, Weierstrass. Of course we recognize these names at once. But a comparable list of American mathematicians from 1776 to 1876 might draw puzzled looks even from an American audience. The list would include Nathaniel Bowditch (1773–1838), best known as author of the *American Practical Navigator* (1802) but who also published a four-volume translation and commentary on Laplace's *Mécanique Céleste*;[1] Theodore W. Strong (1790–1869), who proved some theorems about circles in the early 1800's;[2] Robert Adrain (1775–1843), who published some work on least squares and on the normal law of error,[3] and a host of astronomers, geodesists, surveyors, almanac makers, teachers of mathematics—and one president, Thomas Jefferson, who helped design quite a good mathematics curriculum for the University of Virginia.

To be sure, Benjamin Peirce (1809–1880) of Harvard published a major work in pure mathematics, the *Linear Associative Algebra*, but though he distributed a hundred copies of it in 1870, it was not actually published, and recognized in Europe, until 1881. In fact, in the period before 1876, Peirce was better known for his work in physics, astronomy, and geodesy than for his algebra, and was not yet seen as a towering figure on the purely mathematical scene. J. Willard Gibbs (1839–1903), the American mathematical physicist, is recognized now as the father of vector analysis, but his lectures on vector analysis at Yale did not begin until 1881. In the first hundred years of the republic, then, no American was an outstanding leader in world mathematics.

This bleak situation was widely recognized at the time, both inside and outside the United States. In 1816, for instance, the French philosopher

Auguste Comte was told not to go to the United States because mathematics was not appreciated there; even Lagrange, had he gone there, could have found employment only as a surveyor, Comte was told.[4] In 1840, in a survey of American higher education undertaken for the Corporation of Brown University, the authors lamented, "We have now in the United States...a hundred and twenty colleges....All teach mathematics, but where are our mathematicians?"[5] In 1873, the English mathematician and historian of mathematics Isaac Todhunter observed of the United States that

> with their great population, their abundant wealth, their attention to education, their freedom from civil and religious disabilities, and their success in literature, we might expect the most conspicuous eminence in mathematics.

Nevertheless, he said,

> I maintain that, as against us, their utmost distinction almost vanishes.[6]

Yet somehow, despite these modest beginnings, by the 1890's American mathematics was alive and well. Indeed, it was growing at a furious rate. In a count of American items listed in the German review journal *Jahrbuch über die Fortschritte der Mathematik*, there are four items in 1868, 32 in 1875, 43 in 1877. The number of American articles in one field, algebra, reviewed in the *Jahrbuch* between 1890–1900, is double that for the preceding decade.[7] By the end of the nineteenth century, the work of Americans was known and respected throughout the mathematical world.

The situation I have just described raises the two main questions I shall discuss in this paper. First, why was American mathematics so weak from 1776 to 1876? Second, and much more important, how did what happened from 1776–1876 produce an American mathematics respectable by international standards by the end of the nineteenth century? We will see that the "weakness"—at least as measured by the paucity of great names—coexisted with the active building both of mathematics education and of a mathematical community which reached maturity in the 1890's.

Before we begin answering these questions, let us introduce a useful chronological framework for the period to be discussed. From the Revolution to about 1820, we will find comparatively little mathematical activity. From about 1820 to the 1850's, we will find an interest in improving mathematics education, and much work in applying mathematics to mapping the new continent and the waters off its coast. Finally, from the 1850's to the 1890's, we find a major commitment by an emerging industrial America to a strengthening of the sciences—a strengthening process in which mathematics fully participated.

We will begin by surveying the sciences in general in the nineteenth-century United States. But we will concentrate on what was most important to mathematics.

2. Science in nineteenth-century America. The nineteenth century in general was a great one in the history of science: the century of Faraday and Maxwell, of Helmholtz and Mendeleev, of Darwin and Pasteur. None of these giants of nineteenth-century science was an American. The conditions for scientific research in the United States were relatively poor, for reasons peculiar to American history. First, "knowledge for its own sake" was not much valued in nineteenth-century America, and this remains true throughout our hundred-year period. For instance, in 1832, one American, James Jackson, denied his physician son permission to spend several years studying science before setting up practice. He said:

> We are a business-doing people. We are new. We have, as it were, but just landed on these uncultivated shores; there is a vast deal to be done; and he who will not be doing, must be set down as a drone.[8]

The astronomer Simon Newcomb, as late as 1874, observed:

> However great the knowledge of the subject which may be expected in a professor, he is not for a moment expected to be an original investigator, and the labor of becoming such, so far as his professional position is concerned, is entirely gratuitous. He may thereby add to his reputation in the world, but will scarcely gain a dollar or a hearer at the university.[9]

Because of attitudes like those just described, scientists could often find neither financial support for research nor the time to do it. When physicist Joseph Henry taught at Albany Academy in the 1830's, he taught seven hours a day.[10] As late as the 1880's, twenty hours a week was a common teaching load for professors of mathematics even in major colleges and universities.[11]

Other characteristic American attitudes also worked against science. For instance, after the Revolution, the newly independent Americans at first prized their isolation from Europe—not the best way to be part of a world scientific community. Moreover, science was sometimes seen as anti-democratic; after all, science is done by an elite, not by the common man.[12] Finally, and most important, while a central government in Europe might support science, the United States government was not automatically, by analogy, the patron of science, for federal patronage raised the issue of states' rights. In fact, though the U.S. government did sponsor scientific work when there was an apparent national need to be met, the states' rights issue long served to block the founding of any permanent federal scientific institution in the United States.[13]

Even with these attitudes, however, American natural science from 1776 to 1876 was stronger than its mathematics. The U.S. may point with pride, for instance, to the botanical work of John Torrey and Asa Gray; the physics of Joseph Henry; the founding of the science of oceanography by Matthew Fontaine Maury; and to the existence of a flourishing community of researchers in fields like geology, natural history, astronomy, and meteorology.

The state of mathematics would, then, seem to have been exceptionally low, and we must investigate why whatever promoted the natural sciences in nineteenth-century America did not equally encourage mathematics.

There were, first, two widely shared philosophical attitudes in America in the first half of the nineteenth century which supported the doing of science, if not of mathematics: Natural Theology and the Baconian philosophy.[14] Natural Theology is the doctrine that we may demonstrate the glory of God by discovering the laws of nature; indeed, the existence of natural laws proves the existence of an intelligent creator—God. The doctrine of Natural Theology was part of the world-view of the Puritans, and greatly influenced the colleges of New England in the seventeenth and eighteenth centuries. The doctrine was shared by the Founding Fathers, as reflected in a phrase in the Declaration of Independence: "Nature and Nature's God." Natural Theology, with its religious connotations, was a popular motive for doing science in nineteenth-century America; unfortunately, however, looking for the glory of God in nature is more encouraging to the natural sciences than it is to research in mathematics.

The Baconian philosophy, based on the work of the seventeenth-century philosopher Sir Francis Bacon, stressed three things: first, the importance for science of collecting facts; second, a de-emphasis of, and indeed condemnation of, all-encompassing theories; and, third, the application of science for improving human life. This philosophy was especially congenial to nineteenth-century Americans, both inventors and explorers. In nineteenth-century America, the popularity of the Baconian spirit encouraged the collection of vast amounts of data, especially significant for astronomy and for the biology and geology of a not-yet-explored continent. Baconianism, however, was not especially hospitable to work in mathematics, nor, indeed, to theoretical science in general.

Both Natural Theology and Baconianism were attitudes toward science that the United States had inherited from England. And we should note that another factor discouraging mathematics in the United States was the great influence of English thought, an influence especially marked in the period before 1820. England in the eighteenth and early nineteenth centuries, though producing notable work in the sciences, was quite weak in mathematics. One reason was the English devotion to Newtonian methods— even to notation in the calculus!—methods which by 1800 had been superseded on the Continent by the work of Euler, Lagrange, and Laplace. Another, related reason for England's mathematical weakness was the complete lack of advanced mathematics teaching at English universities.[15] Thus England, which might have served as a source both of inspiration and of textbooks, provided little help to American mathematics.

Nevertheless, there were major forces encouraging American science which also promoted mathematics. A principal impetus for scientific research in nineteenth-century America was the desire to explore, understand,

and subdue the new land. Just as geologists and biologists were needed to learn about the vast continent and its inhabitants, so people knowing mathematics were needed for the exploration: especially to map the coastline and the interior, and to make the astronomical calculations necessary for accurate mapping. National pride contributed also; there was a strong desire not to be dependent on foreign maps and charts for American navigation, for instance,[16] and there was even a proposal made to run the prime meridian through Washington.

The way American needs could encourage mathematical work is illustrated by the career of the first post-revolutionary American mathematical figure, Nathaniel Bowditch of Salem, Massachusetts. Bowditch, a seaman, taught himself mathematics. His original motivation was to understand and improve navigation; one result was his *American Practical Navigator,* first published in 1802 and still being revised and reissued today. Bowditch's major work, however, was a translation and commentary on Laplace's *Mécanique Céleste*; the subject of celestial mechanics was one to which an interest in navigation led many nineteenth-century mathematicians. The commentary was far from trivial; the kind of work involved is illustrated by Bowditch's well-known statement:

> I never come across one of Laplace's *"Thus it plainly appears"* without feeling sure that I have hours of hard work before me to fill up the chasm and find out and show *how* it plainly appears.[17]

For the future of American mathematics, Bowditch's importance includes having engaged the young Benjamin Peirce to help him correct the proofs for the Laplace commentary, thus giving Peirce an introduction to European mathematical physics not then available in any American college.

As Bowditch's interests illustrate, then, the need for accurate maps and charts promoted research in the mathematics related to these tasks: thus error theory, planetary theory, and celestial mechanics benefited. And in these areas, mathematical research was motivated not just by the altruistic desire to fulfill the needs of the nation, but by government funds.

One government institution arising from the need for exploration was the United States Coast Survey. To head the survey, Jefferson—our most mathematical of presidents[18]—found Ferdinand Rudolf Hassler, a Swiss who had worked on the survey of the canton of Bern. Hassler insisted, as a man with European training would, on a sound scientific basis being laid for the work of the Survey.[19] His successors, first Alexander Dallas Bache (1806-1867), and then Benjamin Peirce, shared this strong scientific orientation. The value of science and mathematics for the Survey is clear when we list the Survey's main tasks: the accurate surveying and mapping of the Atlantic Coast and Gulf of Mexico; mapping the Gulf Stream and secondary streams; determining the magnetic force, the depth of the ocean, and the nature of marine life, at various places; and precisely determining longitudes by astronomical observation. The Survey, while doing these tasks, as a consequence performed others as well: it provided its mathematically inclined employees with jobs related to mathematics, with a community to belong to, and with encouragement for work in the mathematics related to geodesy and astronomy. Thus, though the Coast Survey provided no specific impetus for pure mathematical research, it helped build and support a group of professional, mathematically-oriented scientists for the United States.

Another important institution arising from the needs of the new nation was the Nautical Almanac. Founded in 1849, its first head was Navy Lieutenant Charles Henry Davis, both a veteran of the Coast Survey and a former student at Harvard. The major technical problem in compiling a nautical almanac is to determine planetary positions at specific times at pre-determined points.[20] For this, the Nautical Almanac needed good astronomical observations, so Davis set up its headquarters in Cambridge, Massachusetts, which had Harvard College, a good telescope, and Professor Benjamin Peirce. Whatever the Navy may have expected, the atmosphere at the Nautical Almanac office was not so much practical and military as it was academic. Astronomer Simon Newcomb recalled how his going to work at the Nautical Almanac was entering the "world of sweetness and light."[21] Besides Newcomb and Benjamin Peirce, notable American scientists who worked at the Nautical Almanac included philosopher Chauncey Wright; mathematical physicist George William Hill; astronomer Benjamin Apthorp Gould; future M.I.T. president J. D. Runkle; the woman astronomer from Nantucket, Maria Mitchell; and future Wesleyan president J. N. Van Vleck. In 1858, Runkle started a mathematical periodical out of the Nautical Almanac office; called the "Mathematical Monthly," it lasted three years. Thus the Nautical Almanac not only aided the growth of the American mathematical community, it provided that community, at least for a time, with an instrument of communication.

The military needs of the United States provided further occasions for the work of mathematicians, surveyors, and astronomers. For instance, in 1848 Major William H. Emory, a West Point graduate with training in astronomy, led the Mexican Boundary Survey after the U.S. took the southwest from Mexico in the Mexican war. During the Civil War, Superintendent Alexander Dallas Bache made Coast Survey results available to the armed forces of the United States; and accurate Coast Survey data of Virginia in fact played a major part in the capture of Port Royal in 1861 by Captain Samuel F. Du Pont.[22]

And, as the nineteenth century progressed, the applications of the mathematical sciences in America began to extend beyond the calculation of planetary positions and the surveying of the United States. The sciences in general appeared useful both for military purposes and for the growing industries of the United States. The Civil War, for instance, encouraged scientists to study explosives, ironclad ships, the telegraph, and medical statistics. In civilian life, industrialists, particularly those in railroads and textiles, saw a great need for technicians and technically trained managers.[23] Also, the needs of American agriculture for a scientific—and thus more successful—basis were apparent.

Technically trained people were few, however, except for West-Point-trained engineers and the veterans of various surveys. Obviously more scientifically trained civilians were needed, and the place to get them was the educational system. Industrialists and government alike began to provide money for the educational system to produce such technically trained people. (Here developments in the United States parallel those in European industrial nations, especially Germany.) And the improvement of scientific education in the nineteenth century was especially significant for mathematics, since of all the sciences in the nineteenth-century world, mathematics—because it had become so technical—depended most on the educational system to produce competent practitioners. Indeed, outside the educational system it is difficult to become aware that mathematics—as opposed to its applications—can be a career at all. Thus the American educational system was absolutely necessary in generating mathematicians.[23a]

The teaching of mathematics in nineteenth-century America can best be understood by looking at the history of American higher education, especially the trends in science teaching. Accordingly, we will return to the period 1780-1820 to trace that educational history.

3. American higher education and mathematics: to 1850. Most American colleges in the seventeenth and eighteenth centuries had been intended to train ministers; by the early nineteenth century, however, this was no longer their primary function. College education was a gentleman's education, to produce a community of the educated. The educational theory which shaped the early colleges held that there was a pre-existing amount of truth, and that "the primary function of education was to get as much as possible of this corpus of Christian truth into the heads of the undergraduates."[24] If the amount of knowledge is fixed, there is no incentive for research. Another part of prevailing educational theory was that education should provide "mental discipline" for the student. Both the idea of mental discipline and the idea of a fixed body of received knowledge justified the curriculum of these colleges: chiefly, the classics and mathematics; and also some logic, some moral philosophy, and some of what was called "natural philosophy"—physics and astronomy. The scientific part of the curriculum was more oriented toward natural theology than toward practical application; in fact there was really no technical education, save at West Point, much before the middle of the nineteenth century.

Choosing subjects which provided "mental discipline" did give mathematics a greater share of the curriculum than the other sciences. The level of the mathematics taught, however, was not very high, especially when

English influence abounded before 1820. The mathematics usually taught in the colleges was arithmetic, elementary algebra, and the geometry of Euclid, with bits and pieces of surveying, trigonometry, or conic sections thrown in. Even elementary subjects were not always well taught. For instance, an instructor at Dartmouth after the Revolution is supposed to have taught Euclid without proofs, telling his class, "If you doubt the truth of the theorems, read the proofs; but for [my] part [my] mind is satisfied."[25] Even for relatively advanced subjects, teaching was often by rote; in 1830, there was a student riot at Yale against the way mathematics was being taught, known as the "Conic Sections Revolt."[26] Few colleges before the Civil War taught even calculus; almost none required it; and when it was taught the justification might still be "mental discipline." And—most important—the mathematics that was taught was part of a prescribed course of study which left no room for a student to specialize.[27]

Two things were necessary to improve college mathematics teaching: first, departure from the eighteenth-century English model; this was accomplished in the 1820's as part of a general wave of educational reform; second, a new stress on modern science and on a scientific and mathematical curriculum which would meet the needs of a growing industrial society. This second development began at mid-century, but was not really completed until the 1890's. Let us now turn to these two changes.

The old colleges and the education they provided changed because of pressures from American society. The United States, so self-consciously democratic, could not retain the idea of a "gentleman's education" forever.[28] A more practical orientation seemed more relevant to the needs of the nation. And, as the number of colleges multiplied, there was competition between them for students, which encouraged innovation in curriculum. One available innovation was to offer a richer program in science and mathematics.

Both in the United States and in England, educational reformers brought French mathematics into college mathematics teaching in the early nineteenth century. In the United States, the development was encouraged by world affairs. In England after 1812, France was not popular, but in the United States after its War of 1812, it was England that was unpopular. French mathematics came into the United States through the textbooks used to teach mathematics and its applications at French military schools and at the *École polytechnique* in Paris. The *École polytechnique* had been founded during the French Revolution. Education at the *École polytechnique* was intended to produce a *polytechnicien,* one with enough scientific knowledge to be able to apply it to a wide range of problems, and therefore one who knew some mathematics.

In 1818, John Farrar, Professor at Harvard, began a project of translating French mathematics textbooks for American use. Unfortunately, the very latest French mathematics and physics of the 1820's, like that of Cauchy and Ampère, were not included in Farrar's program, but even the eighteenth-century works by men like Bezout, Biot, Lacroix, and Legendre on surveying, trigonometry, algebra, and calculus were a great improvement over what had been taught in the colleges. The French works were both more up to date and more useful for the sciences. Farrar's work at Harvard was not unique; other schools began to teach French mathematics, and other professors undertook translations, notably Elias Loomis at Yale and Charles Davies at West Point. The availability of these new textbooks helped spark a much stronger mathematics program in many colleges.[29]

Just as important as New England colleges like Harvard and Yale in the history of mathematics education in early nineteenth-century America was the work of Sylvanus P. Thayer in reorganizing the curriculum at West Point, on the model of the French military schools and the *École polytechnique*. Besides Thayer, West Point had on its faculty Claude Crozet, a graduate of the *École polytechnique*, and Charles Davies, whose translation of Legendre's introduction to geometry and trigonometry (known as Davies' *Legendre*) was one of the most widely used American mathematics textbooks of the nineteenth century. Because of Thayer, Crozet, and Davies, not only did West Point provide pre-Civil-War America with most of its mathematically trained surveyors and engineers—for instance, providing the Coast Survey with Superintendent Alexander Dallas Bache—but the influence of West Point's mathematical curriculum on American education after the 1830's was immense. The mathematics programs at many schools were developed by professors who were graduates of West Point, including the universities of South Carolina, Mississippi, and Virginia.[30]

By the 1840's, the colleges of the United States still did not provide a scientific education comparable to the best available in Europe. Nevertheless, many American college graduates now had respectable mathematical backgrounds. The colleges may not have produced mathematicians, but they did produce a generation of teachers of mathematics who could respond to the new demands made around 1850 on the American educational system.

4. Mathematics and science education: 1850-1900. By 1850, the railroads, canals, bridges, roads, telegraphs were becoming major factors in the American economy. European research in agricultural chemistry was attracting American attention. Most educated Americans agreed that science was needed to improve industry and agriculture. Furthermore, research, to enlarge the amount of useful scientific knowledge, was also needed. All these things—the growth of industry, the settlement of the continent, and the consciousness of the importance of science—coincided with the growth of great fortunes in nineteenth-century America. Thus private wealth was available to finance education on a scale never before seen in the United States.[31]

In 1847, Abbott Lawrence, the textile magnate, founded the Lawrence

Scientific School at Harvard, hoping to produce graduates able to take their places in modern industry. At about the same time, a scientific school was founded at Yale. In 1861, it was named after Joseph Sheffield, who had funded it with his railroad holdings. The model of European polytechnic institutes—Paris, Dresden, Freiburg—now influenced not only textbooks, but entire institutions; not only the scientific schools at Harvard and Yale, but also (among others) Rensselaer Polytechnic Institute, Brooklyn Polytechnic, and "Boston Tech," later renamed M.I.T.

The land-grant colleges, also, stressed instruction in the sciences. First chartered by the Morrill Act of 1862, the land-grant colleges, which were intended to teach scientific agriculture, eventually included universities like Michigan, Wisconsin, Minnesota, and Cornell, and the various Agricultural and Mechanical schools throughout the nation. The founding of all these schools meant that the sciences in general, and therefore mathematics in particular, were much more widely taught; and the curricula of the new schools influenced the older colleges too. Instruction was not always of the highest quality, since it was impossible to staff so many new schools at once with qualified teachers, and the science which was taught was sometimes narrowly vocational. Nevertheless, the scientific schools provided jobs for mathematicians and scientists, and also made possible a level of instruction that had not existed before.

In the 1850's at the Lawrence Scientific School, mathematics and physics were taught by Benjamin Peirce. Peirce himself had worked under Farrar at Harvard and had assisted Bowditch in preparing his commentary on Laplace, so Peirce's mathematical roots are in the French mathematics of the eighteenth century. But Peirce taught, not eighteenth-century French science, but nineteenth-century European science, including the mathematics and physics of Cauchy, Hamilton, Gauss, and Bessel—without doubt the most advanced mathematical curriculum ever yet seen in the United States. To a man, Peirce's students testify that his lectures, though inspiring, were impossible to follow. Nevertheless, he must have taught them something. His students included many of the most influential scientists and mathematicians of the next generation. For instance, when Cornell in the 1880's developed a mathematics program of international stature, three leading men—Wait, Byerly, and later Oliver—were former students of Peirce.[32] Other Peirce students include astronomers Simon Newcomb, Edward Ellery Hale, and George William Hill; future Harvard Presidents Eliot and Lowell; future M.I.T. president J. D. Runkle; to say nothing of Peirce's sons, Harvard mathematics Professor James Mills Peirce, and the renowned philosopher and logician Charles Sanders Peirce. To be sure, Benjamin Peirce's mathematical curriculum was not home grown; his advanced mathematics and physics were learned from European sources. But his early education, and the financial and institutional support for his work—first French mathematics and physics, then the Nautical Almanac,

Coast Survey, and Lawrence Scientific School—were typical of the situation of mathematicians in the nineteenth-century United States. Benjamin Peirce's career illustrates also the way American mathematics gradually changed from the practical to the more theoretical. Though Peirce was best known in nineteenth-century America for his contributions to applied mathematics, he published, near the end of the century, the first major American contribution to pure mathematics, his *Linear Associative Algebra*. This work, influenced by Hamilton's treatment of quaternions, gave methods of classifying and exhibiting all the linear associative algebras with a given, finite number of fundamental units, making use of the concepts Peirce developed of nilpotent and idempotent elements.[33] And the first sentence of Peirce's work has ever since been quoted as a definition of pure mathematics: "Mathematics is the science that draws necessary conclusions."

A similar pattern may be found in the career of J. Willard Gibbs. His father had been a professor of philology at Yale, and Gibbs grew up in the academic community of New Haven. His education was based on the Yale versions of French mathematics and its applications, and his thesis was in engineering: "On the form of teeth of wheels in spur gearing." After receiving his Ph.D. from Yale in 1863, however, Gibbs went to Germany to pursue his scientific studies, and moved there from engineering to mathematics and physics. He returned to Yale to teach and to do research. But even at Yale, there was no support yet for a great research scientist— literally no support, because Gibbs' professorship carried no salary. Only when Johns Hopkins offered him a job in 1880 did the Yale Corporation arrange that "an annual salary be attached to the chair of mathematical physics."[34] These difficulties notwithstanding, though, advanced scientific training at Yale, as at Harvard, produced students able to teach advanced mathematics and science. Yale's most illustrious mathematics student in the late nineteenth century was future University of Chicago professor E. H. Moore. Moore was at Yale in Gibbs' time, but Moore's major professor at Yale was mathematician Hubert Anson Newton, himself a Yale graduate, and it was Newton who made possible Moore's further education in Germany.

Despite the illustrious careers of Peirce and Gibbs, however, not even the Lawrence and Sheffield schools were dedicated to scientific research. Even after the Civil War, specialized advanced study in the United States generally existed only for those preparing for professions, not for those wanting to pursue knowledge for its own sake.[35] As late as 1875, Charles Sanders Peirce complained that Harvard did not "believe in the possibility of any great advances in science ... being made there," thinking that "the highest thing it can be is a school."[36] But with the intensified American interest in science, this situation could not last long, because in Europe there was a model not only for scientific subject-matter, but a model for the research institution—the German university.

Establishing university education in the United States was almost inevitable by the 1870's. Science-educated students from many American schools, sometimes with European post-graduate study, were available to staff universities; industrial fortunes were available to pay for them; the German model was there to inspire them. A key date is the centennial year 1876, when the first research-oriented university in the United States was funded, from a characteristic source—the fortune in Baltimore and Ohio Railway stock of Johns Hopkins. Johns Hopkins' first president, Daniel Coit Gilman, himself had a science degree from Yale. In fact, the presidents of almost all the research-oriented universities of the late nineteenth century were trained as scientists: F. A. P. Barnard of Columbia, David Starr Jordan of Stanford, A. D. White of Cornell, G. Stanley Hall of Clark, and C. W. Eliot of Harvard.[37] Some universities, like Clark and Chicago, were newly founded in this period; others, like Harvard and Yale, grew out of existing colleges. But whatever the immediate origin of an American university, the sciences were decisive in its development.

The elective system, pioneered by Eliot at Harvard, was part of the new university. The elective system strengthened mathematics in two ways. First, students did not have to study mathematics unless they wanted to, so the way was open for professors to teach more demanding courses. Second, students could deepen their knowledge in a chosen field—mathematics, for instance—as much as they might desire.

Of all the schools I have mentioned, the most important for American mathematics in the 1870's was Johns Hopkins. Because the Test Acts in Britain were not repealed until 1871, the eminent English mathematician J. J. Sylvester, who professed the Jewish religion, was not eligible for a chair at Oxford or Cambridge for much of his career.[38] Since Sylvester was available after his retirement from the Royal Military Academy at Woolwich in 1870, President Gilman made him the first professor of mathematics at Johns Hopkins (not the only time American mathematics has profited from European religious restrictions). Sylvester built a research-oriented department at Hopkins between 1877 and 1883 (before he returned to England to take the Savilian chair at Oxford newly vacated by the death of H. J. S. Smith). Sylvester's Hopkins students went on to teach mathematics and do research all over the United States. Two of them, Fabian Franklin and Thomas Craig, remained at Hopkins; others introduced modern mathematical teaching to many leading American universities: for instance, George B. Halsted at the University of Texas; Washington Irving Stringham at the University of California at Berkeley; C. A. van Velzer at the University of Wisconsin.[39]

Sylvester was not just a teacher and researcher, but the nucleus of an American mathematical community. Accordingly, in cooperation with three Hopkins colleagues, William E. Story, Simon Newcomb, and physicist H. A. Rowland, and with Harvard professor Benjamin Peirce, Sylvester founded the *American Journal of Mathematics* in 1878. Unlike earlier and shorter-lived American journals, the *American Journal* was neither a repository of problems nor an instrument of education; its primary purpose was "the publication of original investigations."[40] Among the articles in the first number of the journal were contributions by American mathematicians at Hopkins, Cincinnati, Princeton, Pennsylvania, and Virginia; two papers from mathematicians in Canada; one from Lipschitz in Bonn; three from Cayley at Cambridge; two from W. K. Clifford in London. Among the first hundred subscribers to the new American journal are, of course, American colleges and the U.S. Coast Survey, but we also find on the list Charles Hermite; the University Library at Cambridge, England; and the library of the *École polytechnique* in Paris—a sort of coming full circle, given the influence of the *École polytechnique* on American mathematical education. American mathematics was clearly on the world map. But Sylvester could not possibly have put it there all by himself, as his unsuccessful stay some thirty-five years before at the University of Virginia shows. Sylvester certainly helped, but, more important, there was by 1880 an American mathematical community, centered at the leading colleges and universities as well as in government agencies.

The level to which American mathematics had reached in the 1880's has been preserved for us by a survey taken for the United States Bureau of Education by Florian Cajori, then Professor at Tulane. The chief mathematics instructional officers in each of 168 colleges and universities answered his questionnaire. Though among the schools not responding were Harvard and Yale, the survey nevertheless provides us with a valuable "stop-action" picture of the change taking place in the United States from the mathematical education of the nineteenth to that of the twentieth century.

Asked, "How many hours do you teach?" many report twenty hours a week, but answers on the order of "ten" appear also. Asked "What else, if anything, do you teach?" 73 of the 118 who answered report teaching subjects outside of mathematics as well as mathematics; of these 73, 32 report teaching outside the physical sciences altogether, including—in 1888!—art, music, bookkeeping, languages, classics, history, and Bible. But 45 of the 118 say they teach only mathematics.[41]

More important, both the quantity and quality of mathematics teaching was increasing. As for the quantity, 112 schools reported that mathematics was elective. When asked, "Is the percentage of students electing higher mathematics increasing?" only three of the 112 report that the percentage is decreasing, 28 say no change, five say "yes" with qualifications, and 76 say without qualification that the percentage is increasing.[42]

The increase in quality can be shown from the type of training now available even to those teachers of mathematics outside the universities, and by the encouragement to do research reported even by college faculty members. Thus, for example, professors at many colleges in 1888 report having studied for a time at institutions providing excellent mathematical education like Johns Hopkins, Harvard, or Yale; and T. H. Safford, Field professor of mathematics and astronomy at Williams College—another former Benjamin Peirce student—reported that his professorship required him "to advance astronomical knowledge."[43]

Finally, in response to the question "What mathematics journals are taken?" 117 of the schools—in 1888!—list *none*. Eleven schools take only the *American Journal*; twelve more take only the *Annals of Mathematics*, which had been founded in 1884 by Ormond Stone (1847-1933) at the University of Virginia. But 28 schools take a number of mathematics journals, including the major European ones.[44] These 28 schools include most major state universities in the midwest and south, the Naval Academy, and private institutions like Northwestern, Vanderbilt, Columbia, and Johns Hopkins.

Thus, though the mathematical standards of Hopkins, Harvard, Yale, and Columbia had not yet trickled down to all schools, the process was well under way. As late as 1904, it is true, 20% of the members of the American Mathematical Society report having studied abroad.[45] But this statistic, paradoxically, helps illustrate the strength of the new American mathematical community. The American university taught these students that European mathematics existed and what it was like; it taught them enough mathematics to benefit from the European training when they got it; and most important, it welcomed them back to use their European training to produce American Ph.D.'s ready to be members of the world mathematical community.

All these trends—economic, educational, and mathematical—came together in the founding by industrialist John D. Rockefeller of the University of Chicago in 1892. Under E. H. Moore, the mathematics department at Chicago became the source of the first generation of American-trained mathematicians of world stature, whose careers will be described in Professor Birkhoff's paper; they included L. E. Dickson, O. Veblen, G. A. Bliss, G. D. Birkhoff, and R. L. Moore. When in 1893 the International Congress of Mathematicians was held under the auspices of the new University at Chicago, invited papers were given not only by illustrious Europeans, but also by thirteen Americans. American mathematics had come of age and was now part of the international mathematical community.

In the 1890's, with the founding of the American Mathematical Society[46] in 1888, the *Bulletin* in 1891, the University of Chicago in 1892, the International Congress in 1893, Felix Klein's Evanston Colloquium of 1893, the *Transactions* in 1900, there was an explosion of mathematical activity in the United States. As we have seen, this explosion in American mathematics was not a creation out of nothing, not a sudden flowering out of previously barren soil. Its roots lie in the influx of French mathematics teaching in the 1820's; it was nurtured by government support for applied mathematics throughout the century, and by the increase in science education which began in the

1850's; and it came to fruition in the universities of the 1870's, 1880's, and 1890's. We may, then, proudly exhibit the institutions and the people that produced the flowering of mathematics in the United States at the end of the nineteenth century as the major achievement of American mathematics in its first hundred years.

Notes

1. On Bowditch, see D. J. Struik, Yankee Science in the Making, New York, 1962, chap. 3, *et passim*; Nathan Reingold, "Nathaniel Bowditch", Dictionary of Scientific Biography, Scribner's, New York, 1970, vol. II, pp. 368-9, with bibliography.

2. On Strong, see George Daniels, American Science in the Age of Jackson, New York and London, 1968, pp. 224-5; cf. H. Poincaré, "Introduction" in The Collected Mathematical Works of George William Hill, Washington, 1905, vol. I, pp. vii-viii, on Strong's influence on his most notable student, G. W. Hill.

3. On Adrain, see J. L. Coolidge, Robert Adrain and the beginnings of American mathematics, Amer. Math. Monthly, 33 (1926) 61-76.

4. Florian Cajori, The Teaching and History of Mathematics in the United States, Washington, 1890, p. 94. This book is a gold mine of all sorts of information.

5. Richard Hofstadter and C. De Witt Hardy, The Development and Scope of Higher Education in the United States, New York and London, 1952, pp. 24-5.

6. Cajori, *op. cit.*, p. 99.

7. D. E. Smith and Jekuthiel Ginsburg, A History of Mathematics in America before 1900, Chicago, 1934, 154-157.

8. Quoted by Donald Fleming, William H. Welch and the Rise of Modern Medicine, Boston, 1954, p. 8.

9. I. B. Cohen, Science in America: The Nineteenth Century, in A. M. Schlesinger, Jr., and Morton White, Paths of American Thought, Boston, 1963, pp. 167-189; Newcomb's statement is quoted on p. 185.

10. Nathan Reingold, ed., Science in Nineteenth-Century America: A Documentary History, London, Melbourne, Toronto, 1966, p. 71.

11. Cajori, *op. cit.*, pp. 345-9.

12. Nathan Reingold, American Indifference to Basic Research: A Reappraisal, in George Daniels, ed., Nineteenth-Century American Science: A Reappraisal, Evanston, Ill., 1972, pp. 38-62, see p. 60; cf. Struik, *op. cit.*, p. 239.

13. See A. H. Dupree, Science in the Federal Government, Cambridge, Mass., 1957, p. 5 *et passim*; F. Rudolph, The American College and University: A History, New York, 1962, records opposition on this basis even to the land-grant colleges, p. 250.

14. Daniels, American Science in the Age of Jackson, Chapters 3-5.

15. W. W. R. Ball, A History of the Study of Mathematics at Cambridge, Cambridge, 1889.

16. Struik, *op. cit.*, p. 412.

17. Cajori, *op. cit.*, p. 104 (italics in original).

18. D. E. Smith, Thomas Jefferson and mathematics, Scripta Math., 1 (1932-33) 3-14.

19. Dupree, *op. cit.*, pp. 53-55; Struik, *op. cit.*, pp. 406-7.

20. Simon Newcomb, The Reminiscences of an Astronomer, Cambridge, Mass., 1903, pp. 63-4.

21. Newcomb, *op. cit.*, Chapter 3.

22. Dupree, *op. cit.*, p. 133.

23. H. Miller, Dollars for Research: Science and its Patrons in Nineteenth-Century America, Seattle and London, 1970, p. 75; B. Sinclair, The Promise of the Future: Technical Education, pp. 249-72 in G. Daniels, ed., Nineteenth-Century American Science: A Reappraisal, p. 261; cf. Hofstadter-Hardy, *op. cit.*, p. 31.

23a. R. V. Bruce, A Statistical Profile of American Scientists, 1846-1876, pp. 63-94 in Daniels, Nineteenth-Century American Science, pp. 87-9, shows that, among American scientists in the mid-nineteenth century, mathematicians were 50% more likely than other scientists to have chosen their field because of their experiences at school. In career choices made by other scientists, family influences or possible jobs were more significant.

24. Hofstadter-Hardy, *op. cit.*, p. 14; cf. the Yale Report of 1828, summarized in Rudolph, *op. cit.*, pp. 130-5.

25. Cajori, *op. cit.*, p. 74.

26. Cajori, *op. cit.*, p. 153; Stanley Guralnick, Science and the Ante-Bellum American College, Philadelphia, 1975, p. 55.

27. Daniel Kevles, On the Flaws of American Physics: A Social and Institutional Analysis, pp. 133-151 in Daniels, Nineteenth-Century American Science, points out that this was also a factor discouraging the growth of American physics, p. 136.

28. Hofstadter-Hardy, *op. cit.*, p. 22; Rudolph, *op. cit.*, Chapters 6, 10.

29. Stanley Guralnick, *op. cit.*, chapter 3; L. G. Simons, The influence of French mathematics at the end of the eighteenth century upon the teaching of mathematics in American colleges, Isis., 15 (1931) 104-23.

30. Cajori, *op. cit.*, pp. 195, 209, 220, 248.

31. Hofstadter-Hardy, *op. cit.*, p. 31.

32. Cajori, *op. cit.*, p. 178.

33. Benjamin Peirce, Linear associative algebra, Amer. J. Math., 4 (1881) 97-215; addenda, pp. 216-229. For the place of this work in the history of algebra, a brief account may be consulted in E. T. Bell, The Development of Mathematics, New York, 1945, esp. p. 249. On Peirce, see first Carolyn Eisele, "Benjamin Peirce", Dictionary of Scientific Biography, vol. 10, pp. 478-81, with bibliography.

34. Reingold, *op. cit.*, pp. 318-19.

35. Hofstadter-Hardy, *op. cit.*, p. 61.

36. Reingold, *op. cit.*, p. 228.

37. Hofstadter-Hardy, *op. cit.*, p. 33.

38. For an account of the nine-year campaign in Parliament to repeal the Test Acts, as well as some examples of their effects at Oxford and Cambridge, see D. A. Winstanley, Later Victorian Cambridge, Cambridge, England, 1947, chapter 3. These laws were, to be sure, anachronisms in the society of late nineteenth-century Britain, and Sylvester was showered with honors from his countrymen, from being an FRS at age 25 to his presidency of the Mathematics and Physics section of the British Association for the Advancement of Science in 1869. Nevertheless, he officially could not even take his degree at Cambridge until 1871.

39. Cajori, *op. cit.*, p. 272. Stringham, by the way, had been an undergraduate at Harvard and studied under Benjamin Peirce. Halsted, before going to Texas, also taught for five years at Princeton.

40. Amer. J. Math., Pure and Applied, I, p. iii.

41. Cajori, *op. cit.*, pp. 345-9.

42. *Ibid.*, p. 303.

43. *Ibid.*, pp. 346-9; p. 347, 386.

44. *Ibid.*, p. 302.

45. D. E. Smith and J. Ginsburg, *op. cit.*, p. 112; the statistics come from T. S. Fiske's presidential address to the American Mathematical Society in 1904.

46. First called the New York Mathematical Society. For the history, see R. C. Archibald, A Semicentennial History of the American Mathematical Society, 1888-1938, in American Mathematical Society: Semicentennial Publications, vol. I, New York, 1938.

SOME LEADERS IN AMERICAN MATHEMATICS: 1891–1941

Garrett Birkhoff

1. Preface. On this bicentennial celebration of the founding of our nation, I am happy to pay tribute to the men most responsible for making our nation a world leader in mathematics. Some of these men have already told their own story in the two AMS Semicentennial volumes ([1], [2], [5]), covering a period 1888-1938 nearly identical with the half-century 1891-1941 that I shall be discussing. What I can add to their first-hand accounts is primarily the perspective of a mathematician who was beginning his career at the end of this period, and who had the good fortune to know many leading American mathematicians of the period personally.

I can also supply a few personal recollections of my father, George D. Birkhoff (1884-1944). Since a thorough and perceptive appraisal of G. D. Birkhoff's scientific work has been made by Marston Morse [6, vol i, pp. xxii-1vi], I shall emphasize here some human aspects of his career, insofar as I have been able to reconstruct it from our casual discussions, in the rosy light of hindsight. I have prepared these remarks with very special care and affection, as I do not plan to publish any other reminiscences about G. D. Birkhoff; I am sure he never intended me to be his biographer!

Finally, I can recall some impressions of the Harvard community in which I have lived most of my life, as it was in my youth. I have often been asked to do this, and I hope that the disproportionate space that I have devoted to these impressions will not be misconstrued as an evaluation of Harvard's importance for American mathematics in that half-century, great as it undoubtedly was.

American mathematics in 1891. Having made these apologies, let us recall briefly the state of American mathematics in 1891. Judith Grabiner has brought out clearly, if perhaps too humbly, the paucity of our nation's contributions to mathematics during the first century of its existence. She has also described the beginnings of organized research during the period 1876-1891. Two mathematical research journals were founded during these years: the *American Journal of Mathematics* at the Johns Hopkins University (JHU), under the editorship of J. J. Sylvester and Simon Newcomb; and the *Annals of Mathematics,* under the editorship of Ormond Stone at the University of Virginia.

These journals and their influence were also analysed by Thomas Fiske (1865-1944), a founder of the New York Mathematical Society (NYMS) in 1888, in his presidential address to the American Mathematical Society (AMS) in 1904.* He notes that of the 90 contributors to the first ten volumes of the *American Journal of Mathematics,* thirty were foreign and "almost one-third of the remaining sixty were pupils of Professor Sylvester". Thus the distinguished research at the JHU in those years can hardly be regarded as indigenous!

In the 1890's, the leadership of American mathematics was firmly centered in New York. Here the NYMS, soon to become the AMS, was burgeoning. It had 135 members in May, 1890; by year's end it had 210. These were not all pure mathematicians; forty per cent applied mathematics professionally to astronomy, physics, engineering, or actuarial work. In 1891, these men founded the *Bulletin of the New York Mathematical Society: A Historical and Critical Review of Mathematical Science.* The *Bulletin* changed NYMS to AMS when our Society changed its name, but its nature did not change; and its official description as a "review" remained on the cover until 1931. My narrative therefore begins with the founding of the *Bulletin.*

The lack of emphasis in pure mathematics in nineteenth century America is not surprising. As President Lowell of Harvard (see §12) said in 1930:† "The great task of the United States in the nineteenth century was filling a continent and creating its industries. That of the twentieth should be raising its civilization to the highest level attainable." This level was, of course, exemplified by Western Europe at that time. As a result, the ambition to equal European culture was widespread and natural in our country. It was felt not only by university people, but also by wealthy philanthropists like John Nicholas Brown, Johns Hopkins, Daniel Marsh Rice, Leland Stanford, Andrew Carnegie, and Cornelius Vanderbilt, who made princely gifts to strengthen the great universities which perpetuate their names.

By 1891, most of the greatest leaders in American mathematics of the pre-1941 era were already born. In the years of their boyhood, most Americans put in long hours of hard physical labor to produce plentiful food, clothing, and housing. Automobiles were a curiosity, airplanes were un-

---

*Bull. AMS 11 (1904) 238-46.
†A. L. Lowell, "At War with Academic Traditions in America," Harvard University Press, 1934, p. 332.

known. Travel, a leisurely affair by railroad, steamship, and horse drawn carriage, was the privilege of the few. Communication was mostly by letter; even telephone service was a novelty. News was transmitted telegraphically; the radio was non-existent, and TV undreamed of. Though all these things were to become commonplace during the lifetimes of the men I shall talk about, our nation was very much a "developing country" of 60 million inhabitants in 1891. It had high literacy and many colleges, but very few graduate schools, and these had existed for only a few decades. Moreover, American culture was still regarded as quite provincial by most Europeans.

However, the stage was set for the dramatic rise of American mathematics to world leadership by 1941, and this rise will be my main theme. In 1891, most of the greatest leaders of the AMS during the decades 1921–1941 were still boys. By the time they reached manhood, they felt the challenge to emulate the best in European mathematics, and they successfully met this challenge. The rest of my talk will be concerned with their personalities and accomplishments. I can only hope that the young mathematicians of today will anticipate similarly the challenges of the next fifty years, and meet them as successfully!

PROGRESS BEFORE WORLD WAR I

2. The first AMS presidents.‡ As I have already said, of the 200-odd NYMS members listed in the *Bulletin* of November, 1891, forty per cent had at least partial professional responsibility for astronomy, physics, engineering, or actuarial work. What is more striking, *all five* presidents of the NYMS-AMS in the 19th century were applied mathematicians or administrators!

Thus its first president was a popular college administrator (J. H. van Amringe, 1835-1916), whose early papers had dealt with life insurance. Its second president (John E. McClintock, 1840-1916) was a prominent actuary. Its third and fourth presidents were the notable mathematical astronomers G. W. Hill and Simon Newcomb, about whom I shall say more shortly. The fifth was R. S. Woodward (1849-1921), who like Gauss was an astronomer, surveyor, and mathematical physicist. He was rated twenty-first among American mathematicians and eleventh among American physicists in 1903. After being at Columbia from 1893 to 1904, he became the first President of the Carnegie Institution of Washington, serving as such until his death at age 72.

From Fiske's account [1, pp. 6–7], one gets the impression that the AMS

---

‡For excellent short biographies of all AMS presidents to 1938, see [1, pp. 107-243]; their own attitudes are well expressed in their presidential addresses, published in the Bull. AMS. For Fiske's biography of McClintock, see Bull. AMS 23 (1917) 353-7.

meetings over which these men presided were convivial affairs. Osgood likewise, in describing the planning meetings for the *Transactions* [1, p. 58] mentions a *Rathskeller* atmosphere. And Wiener, recalling a 1915 AMS meeting in [19, p. 225], writes of "a beer-hall flavor, which has evaporated with time and the increased prosperity and respectablity of the scientist."

The beery conviviality that pervaded early NYMS and AMS meetings was however misleading, at least as regards its presiding officers. Hill and Newcomb were regarded in Europe as the leading American mathematicians of their time. Both were invited to address International Congresses, whereas, E. H. Moore† and Osgood were passed over. It was not until 1912 that an American pure mathematician (Maxime Bôcher) achieved this honor. It seems in order, therefore, to say something about the careers of these two astronomer-presidents of the AMS, and of the importance of celestial mechanics for the mathematics of their time.

Hill.‡ George William Hill (1838-1914) was a lifelong bachelor; he went to Rutgers but never did any formal graduate work in mathematics. He earned his living from 1861 to 1892 by working on the Nautical Almanac (shades of Bowditch!), living much of the time in solitude on his farm in New Jersey. Though he lectured on celestial mechanics at Columbia several times, he was never a college professor in the ordinary sense. He became famous for his highly original use of infinite determinants to establish the existence of periodic orbits (the Hill equation), and of analytic continuation to perturb them. Poincaré rigorized and generalized Hill's methods, which foreshadowed the functional analysis of Volterra and Fredholm, in his *Méthodes Nouvelles de la Mécanique Céleste*.

Newcomb. Though Simon Newcomb (1835-1909) had even less formal education than Hill, and never did anything equally fundamental, he was a much greater worldly success. After turning down the Directorship of the Harvard Observatory, he accepted a professorship at the Johns Hopkins University, where he became the dominant editor of the *American Journal* after Sylvester's return to England. Though Newcomb ranked Hill as "easily the greatest master of mathematical astronomy during the last quarter of the nineteenth century" he also thought of him as an employee whose annual salary he managed with great difficulty to get increased from $1200 to $1400 [16a, pp. 218-24].

A communicative extrovert, Newcomb was showered with honors through-

---

† Moore spoke in 1908 and 1912 (see §3), but never by invitation.

‡ See [25], vol. 8, pp. 275-312 and Bull. AMS 21 (1915) 499-511 for biographies by E. W. Brown. Also Amer. Jour. Math. 60 (1933) #4, which was dedicated to Hill.

out the latter part of his life. Just to list them takes two pages of fine print![‡]
More representative of his personality, which was that of a natural philosopher, were his 146 non-astronomical papers on every conceivable subject—
he almost became a political economist. He combined immense astronomical
calculations with some experimental work, even collaborating with Michelson on a fresh determination of the velocity of light. Newcomb was also
the grandfather of Hassler Whitney (see §20).

Celestial mechanics. The influence of celestial mechanics on mathematics in the pre-1914 era was very great. It was esteemed not only for its
scientific interest, but also because of its practical importance for navigation and the tides, already mentioned by Professor Struik. It was this
that led to the establishment of the *Nautical Almanac* office in Cambridge,
whose influence on American mathematics Judith Grabiner has described.
Even earlier, it may well have been the stimulus of proofreading Laplace's
Mécanique Céleste for Nathaniel Bowditch that first attracted Benjamin
Peirce to mathematical research. Likewise, the founder of the *Annals of
Mathematics* was an astronomer, Ormond Stone; he remained an Editor
until 1924 and an Associate Editor through 1932. As I have just said, the
third and fourth NYMS-AMS presidents were the notable mathematical
astronomers G. W. Hill and Simon Newcomb. Another astronomer-
mathematician active in AMS affairs was John M. van Vleck (1833–1912),
three of whose mathematical students (E. B. van Vleck, H. S. White, F. S.
Woods) gave the 1903 Colloquium Lectures. I shall mention still others
in §8.

Fiske. Thomas Scott Fiske (1865-1944), the founder and seventh president of the AMS, personified the transition from the AMS of the 1890's
to what it rapidly became. Like most of the other AMS presidents I have
discussed, he was primarily an administrator; from 1901 to 1936, he acted
as secretary (and later treasurer) of the College Entrance Examination
Board. But unlike them, he had a Ph.D. (from Columbia), and was interested primarily in *pure* mathematics.

3. E. H. Moore and Chicago. After 1900, the AMS was dominated by
a new generation of leaders, primarily oriented towards pure research.
Chief among these were E. H. Moore, Osgood, Bôcher, and Fine, all of
whom had gone to Germany in the 1880's to get the training for scientific
research work which German universities offered. They had all returned by

---

‡[1, pp. 124–6]. Fuller biographical sketches have been given by E. W. Brown, Bull. AMS 16
(1910) 341–55, and by W. W. Campbell in the Memoirs Nat. Acad. Sci. 17 (1924) 1–18;
Newcomb's bibliography (compiled by R. C. Archibald, a fellow Nova Scotian) continues
through pp. 23–67.

1891; the oldest (Fine) was then 33, and the youngest (Bôcher) just 24. They aspired to establish in our country traditions of mathematical research similar to those in Germany. The achievements of these four leaders and their students during the years 1891-1914 will constitute the main theme of the next part of my talk.

In the late 1890's, Moore worked with Bôcher, Osgood, James Pierpont of Yale, Maschke, and T. S. Fiske to provide the AMS with a research journal of its own. After efforts to persuade Simon Newcomb to relinquish his autocratic control over the *American Journal of Mathematics* failed,* the AMS founded its *Transactions* in 1900 with Moore, Fiske, and E. W. Brown as its first editors. Its aim was to supply a more adequate publication outlet for American mathematical research.

The establishment of the *Transactions* as a national research journal in 1900 set the stage for a new level of mathematical activity. Much of this was stimulated by Moore, Osgood, Bôcher, and Fine, who took turns at being AMS president during eight of the next twelve years—Fiske and H. S. White serving the other four.

The most remarkable of these leaders was E. H. Moore (1862-1932)†. E. T. Bell was exaggerating only slightly when he wrote [1, p. 3]: "In the late 1890's and early 1900's the history of mathematics in this country was largely an echo of Moore's success and enthusiasm at the University of Chicago ..."

Already as a high-school student, Moore had become fascinated by mathematics while working at the Cincinnati Observatory as emergency summer assistant to Ormond Stone, the founder of the *Annals of Mathematics*. He then went to Yale, which had set up a Graduate School in 1842. Here he got a Ph.D. in 1885, after six years as an undergraduate and graduate student. His thesis (on n-dimensional geometry) was published in the Trans. Conn. Math. Sci (1885) 9–26, in which Gibbs had also published his most important work. After a summer in Göttingen, where he met Felix Klein and other German mathematicians, he spent a year at the University of Berlin, where Weierstrass (then 70) and especially Kronecker inspired him.

When the University of Chicago opened in 1892, thanks to a munificent gift from John D. Rockefeller, Moore became the first chairman of its mathematics department. He had already secured professorships for Heinrich Maschke (1853-1908) and Oskar Bolza (1857-1942), two gifted mathematicians who had studied in Berlin and acquired Ph.D.'s under Felix Klein in Göttingen.

Felix Klein. Felix Klein's influence on the development of American

---

*See [1, Ch. V] for some details about these negotiations.

†For Moore's career, see [1, pp. 144–50]; G. A. Bliss, Bull. AMS 39 (1933) 831–8, and 40 (1934) 501–14; also [25, vol. 17, 83–102].

mathematics was so great that it seems appropriate to recall a few facts about him. Born in 1849, he got his Ph.D. when only 19. His inaugural professorial address at Erlangen, delivered when he was 23, was "perhaps the most influential and widely read mathematical lecture of the last 60 years."† It showed how the principal branches of geometry (Euclidean, affine, projective, conformal, etc.) are characterized by the *groups* leaving their respective fundamental concepts invariant.

Early in 1893, the year of the Chicago World's Fair, Moore, Maschke, Bolza, and H. S. White (also a Göttingen Ph.D. and newly arrived at Northwestern) sent out invitations for an International Mathematical Congress to be held in Chicago that summer. Mathematicians from six European countries participated, most notably Felix Klein, who gave a series of lectures after the Congress at Northwestern University, later published by the AMS. Klein's superb philosophical and historical perspective makes them good reading even today.‡

Thus Chicago became overnight *the* leading center of American research, with Moore as its inspired chief. In 1858, Kronecker had shown that every finite Abelian group was a direct product of cyclic groups of prime-power order.* In 1893, Moore proved the analogous but deeper result that every finite field is a Galois field: the root field of $x^q = x$ for some prime-power $q = p^r$. During the 1890's Moore also wrote on Steiner's triple systems and other combinatorial configurations, as well as on finite groups, stimulating Dickson's early work on these subjects (see §5).

The year 1899 had witnessed the appearance of the first edition of the *Grundlagen der Geometrie* by Moore's contemporary David Hilbert (1892-1931). This book contained the first fully rigorous discussion of the axioms of Euclidean geometry. It stimulated E.H. Moore and his student Veblen (see §6) to try to improve on these axioms. This they did in articles appearing in the *Transactions* in 1902 and 1904; R. L. Moore extended their work some years later (see §10).

Around 1905, Moore became fascinated by the rapidly burgeoning theory of integral equations, to which Hilbert was making major contributions. After the publication of Fréchet's Thesis in 1906, and under the spell of Peano's symbolic approach to mathematics, he tried to lay new and extremely general foundations for what would today be called functional analysis, but which he called 'general analysis.' He first presented his ideas in his 1906 New Haven Colloquium Lectures. Here he emphasized his famous principle of generalization by abstraction: "The existence of analogies between central features of various theories implies the existence of a

---

†R. Courant, Jahresb. Deutsche Math. Ver. 34 (1926) 195-213.
‡For the 1893 Congress and subsequent lectures, see [1, Ch. VI], and the 1893 NYMS *Bulletin*.
*See also Gauss, *Werke*, vol. 2, p. 266, and E. Schering, Gott. Abh. 14 (1869) 3.

general abstract theory which underlies the particular theories and unifies them with respect to those central features."

Moore presented his ideas again at the International Congresses of 1908 (in Rome) and 1912 (in Cambridge, England). Though very original, they had little influence at the time, probably because they were so abstract and often couched in Peanese symbolism. Indeed, not many of Moore's innovative ideas about General Analysis have survived.‡ Probably most important are those of 'extensionally attainable property', 'directed set', and 'relative uniform convergence,' whose significance was finally appreciated in the 1930's when abstract mathematics became fashionable (see §20).

Even more notable than Moore's own research achievements were those of his Ph.D. students. These included L. E. Dickson, Oswald Veblen, G. D. Birkhoff, T. H. Hildebrandt, and 25 others. R. L. Moore, who got his Ph.D. at Chicago in 1905, must also have been inspired by his namesake. Moreover Wedderburn's great masterpiece (§6) followed his year at Chicago; while Mac Lane mentions that as late as 1930-31, his interest in algebra was aroused by Moore's lectures.

Although Moore's lectures were exciting for good mathematicians, they often confused students not sharing his enthusiasm for mathematics. In this tendency, he resembled Benjamin Peirce and many other research-oriented mathematicians.

4. Harvard: Cole, Osgood, and Bôcher. While E. H. Moore was inspiring an outstanding group of algebraists and geometers at Chicago, a renaissance of activity in classical analysis was taking place at Harvard. In 1872, after previous faltering efforts, President Eliot established a Graduate Department with Benjamin Peirce as Dean. This was four years before Johns Hopkins got started. Coolidge [3] credits Benjamin's son, James Mills Peirce (A.B. 1853) with also "fostering the Graduate School in its early years." Benjamin's other son Charles Sanders Peirce, though far more original, was too reckless to become a professor (see[1, p. 6] for some amusing anecdotes).

The first Ph.D. of Harvard's Graduate School was granted in 1872 to William Elwood Byerly, a distinguished analyst whose course on Fourier series and boundary value problems introduced Harvard students to mathematical physics for decades. Closely associated with Byerly was Benjamin Osgood Peirce (A.B. 1876), a distant relative of the other Peirces. B. O. Peirce was an active researcher who kept mathematics and physics well-coordinated at Harvard for many years.† In 1888, Ginn published a text by Peirce on the "Theory of the Newtonian Potential Function"; Byerly's

---

‡For a sympathetic appraisal of General Analysis, see O. Bolza, Jahresb. DMV 23 (1914) 248-303. For the applications to linear integral equations, see E. H. Moore, Bull. AMS 18 (1912) 334-62.

†See his biography by E. H. Hall in [25, vol. 8, 437-68].

"Fourier Series and Spherical Harmonics" followed in 1893. These were the texts for Math. 10a and 10b, courses that were still given when I was in college.‡ Peirce's *Table of Integrals*, first published in 1899, is still in use, as revised by Osgood and later by Foster.

F. N. Cole. Not all the distinguished mathematicians educated at Harvard during that period were analysts. An early algebraist from Harvard was Frank Nelson Cole (1861-1926). After graduating in 1882, Cole was awarded a Parker Fellowship for study in Germany, like Osgood, Bôcher, Hedrick, and many others after him. There he got a Ph.D. with Felix Klein in 1886, returning to lecture at Harvard for three years before going to Michigan. While at Michigan, he pioneered in introducing the study of finite groups in our country (see §5). In 1896, he went to Columbia, where he served as AMS Secretary until 1921, and chief editor of the *Bulletin* from 1899 to 1920. Cole's personality and his influence on the AMS for a quarter of a century are vividly described in [1, pp. 100–3]; the Cole Prize and vol. 27 of the *Bulletin* are permanent tributes to his memory.

Osgood and Bôcher.† Although Byerly, the two Peirces, and F. N. Cole all contributed significantly to the development of American mathematics, Harvard's national leadership and international reputation during the decades 1894-1914 were primarily due to William Fogg Osgood (1864-1943) and Maxime Bôcher (1867-1918). After graduating from Harvard (Osgood in 1886, Bôcher in 1888), both men got Parker Fellowships for study in Germany, where they got Ph.D.'s (Osgood at Erlangen in 1890 at age 26, and Bôcher at Göttingen in 1891 while still only 23); moreover, both men were influenced by Felix Klein.

Their careers were similar in other respects. Klein invited them both (but apparently not Moore) to write articles for the *Encyclopädie der Mathematische Wissenschaften*, of which he was the chief organizer. Finally, both became Presidents of the American Mathematical Society, Osgood during 1905-6 and Bôcher during 1909-10.

However, their scientific personalities were very different. Osgood was rigorous, systematic, and thorough. He expanded his article on complex anaylsis in the *Enc. Math. Wiss.* into a treatise, *Lehrbuch der Funktionentheorie* (1907), that was preeminent for at least 20 years. His own theorems in this and other areas of analysis were notable for their sharpness and

---

‡By that time, Kellogg's *Potential Theory* (still the best book on the subject today) had replaced Byerly's text. Kellogg was a Ph.D. of Hilbert.

†For Osgood, see [1, pp. 153–8]; and B. O. Koopman, Bull. AMS 50 (1944) 139–42. Osgood's imitation of Felix Klein's manner is ridiculed by Wiener in [22, pp. 231–3] and [23, p. 30]. For Bôcher, see [1, pp. 161–6] and G. D. Birkhoff's appreciation in Bull. AMS 25 (1919) 197-215 (reprinted in [5, vol. iii, pp. 227–45]); and Osgood in Bull. AMS 25 (1919) 337-50.

Weierstrassian rigor. He taught at Harvard from 1890 to 1933, often from his own widely used textbooks, and his course on functions of a complex variable remained *the* key course for Harvard graduate students until World War II.

Bôcher, on the other hand, was intuitive, brilliant, and fluent. His short monographs on integral equations and on the methods of Sturm were models of lucidity and perception, as was his invited address at the 1912 International Mathematical Congress. Whereas Osgood supervised only four Ph.D. Theses in his 43 years at Harvard, Bôcher supervised seventeen in 24 years. His students included D. R. Curtiss, G. C. Evans, Lester Ford, Tomlinson Fort, and E. R. Hedrick (see §9).

Both Osgood and Bôcher wrote landmark texts for American college students, which played an important role in my own education. Thus, I learned the calculus from Osgood's books on the subject, and he was co-author of the book from which I learned analytic geometry (he also wrote one on mechanics). Bôcher was co-author with my high-school teacher Harry Gaylord of a text on trigonometry, but it was his *Introduction to Higher Algebra* (1907), later translated into German and Russian, that was most famous. When Saunders Mac Lane and I were writing our *Survey of Modern Algebra,* I had this book and H. B. Fine's *College Algebra* much in mind, as the books whose substance we should reformulate axiomatically before emphasizing the general theories of (abstract) groups, rings, and fields.

5. Dickson and Bliss. By 1900, the University of Chicago had already turned out two outstanding new mathematicians: Leonard Eugene Dickson (1874-1954) and Gilbert Ames Bliss (1876-1961). I shall next sketch the careers of these two men, up to the time when they became AMS presidents (Dickson in 1917–18, Bliss in 1921–23). During these years, Dickson was much more influential than Bliss, and his activities will therefore dominate this section.

Dickson was a native Texan, who had much of the dynamic energy and rugged individualism that we associate with that state. During his lifetime, he wrote nearly 300 papers and 18 books (for a list through 1937, see [1, pp. 183-94]), and supervised at least 64 Ph.D. Theses. Dickson had already published 10 papers before getting his Ph.D. with E. H. Moore in 1896, at the age of 22. During the next decade, Moore's influence[†] was reflected in Dickson's concentration on finite groups (e.g., the simple groups of order $2^6 3^2 5$), finite fields (see Bull. AMS 6 (1900) 203-4) and "tactical configurations". In 1900, he joined the Chicago faculty; in 1903,

---

[†]In turn, Moore's and Cole's interest in groups may have been stimulated by Felix Klein, a great apostle of the subject (Lie and Frobenius were among his converts).

he was ranked as the ninth best American mathematician; and by 1906, he had already published 126 papers!

Dickson's interest in finite groups was doubtless stimulated not only by Moore, but also by F. N. Cole (see §4), who had translated Netto's *Theory of Substitutions* into English in 1892, and then written seven more papers in the next four years concerned with the enumeration and general properties of finite abstract groups, permutation groups, and linear groups of specified order and degree.

Linear algebras. Another phase of Dickson's early work was concerned with linear associative algebras, as defined by Benjamin Peirce. Consider-able insight into the original stimuli for this work is gained by reading a contemporary paper by Wedderburn,* who came to Chicago from Scotland in 1904-5 as a Carnegie Fellow. In this paper, Wedderburn sharpened E. H. Moore's result that every finite field is a Galois field, by showing that the hypothesis of commutativity is redundant. Since every finite division ring is a division algebra over some Z_p, it suffices to show that every finite division algebra is a field. Wedderburn proved this, using a result from number theory announced by G. D. Birkhoff and Vandiver in 1902, when G. D. Birkhoff was only 18! The comments at the end of Wedderburn's paper also reveal the influence of Moore and Dickson!

Dickson's own work on linear algebras was primarily concerned with division algebras (see Trans. AMS 7 (1906) 370-90 and 514-22) and es-pecially with cyclic division algebras. He continued to contribute to the theory of linear algebras for many years, and his approaches were developed further in the early 1930's by A. A. Albert, his best student. However, his later work dealt increasingly with number theory (see §13).

G. A. Miller.† The interest of Cole, Moore, and Dickson in finite groups was contagious! Among those attracted to the subject, one of the most enthusiastic was G. A. Miller (1863-1951), who fell in love with the subject while living in F. N. Cole's home at Ann Arbor. Miller had begun his career by correcting and extending the lists of transitive and primitive permutation groups of low degree worked out by Serret (1850), Jordan (1872) and Cayley (1891). Since American research on finite groups to 1938 has been exhaustively (if not always accurately)‡ reviewed by E. T.

---

*See J. H. Maclagan-Wedderburn, Trans. AMS 6 (1905) 349-52, and G. D. Birkhoff and H. S. Vandiver, Annals of Math. 5 (1904) 173-80.

†For Miller's biography, by H. R. Brahana, see [21], vol. 30, pp. 257-76 (plus 35 more pages of bibliography).

‡E.g., his mention of the "simplicity" (should be "solvability") of groups of order $p^\alpha q^\beta$.

Bell in [2, pp. 8-15], I shall only add that Miller, Blichfeldt, and Dickson collaborated in writing the book *Finite Groups* (Wiley, 1916) which introduced me to group theory 15 years later, and that Miller bequeathed two million dollars to the University of Illinois, where he had taught for many years.

Maschke and Bolza. Although Dickson and Veblen were primarily influenced by Moore, the outstanding success of the University of Chicago in turning out first-rate mathematicians also owed much to Maschke and Bolza. Moreover Maschke played an important role in the founding of the *Transactions* [1, pp. 57-9],* though his influence on students seems to have been much less than that of Bolza.

Bolza, after studying with Christoffel, Weierstrass, and H. A. Schwarz, got his Ph.D. with Felix Klein in 1886. His *Lectures on the Calculus of Variations* (1904) are still a superb reference, while in a 1914 paper he defined and made the first serious attack on the Problem of Bolza: that of minimizing a definite integral on a curve whose endpoints are constrained to lie on given surfaces. When Bolza returned to Germany in 1910, after Maschke's death in 1908, the University of Chicago had truly suffered "an irreparable loss" [1, p. 78].

Bliss. Bliss got his Ph.D. in 1900; he was an analyst whose main mathematical inspiration came from Bolza, and his most important mathematical work was in Bolza's field: the calculus of variations. This interest was intensified by a year spent at Göttingen with Klein, Hilbert, Minkowski, Zermelo, Carátheodory, and others. Though not as original or prolific as Dickson, he was (like his mentor Bolza), an excellent scholar and a masterful expositor and interpreter.

By 1908, he had published 16 research papers, and been invited (with E. Kasner of Columbia) to give the 1909 colloquium lectures. These were the basis of his monograph on *Fundamental Existence Theorems* (Amer. Math. Soc., 1913), an invaluable reference for a generation of American graduate students.

It was natural that, when Maschke died, Bliss should be asked to take his place. Dickson had been on the Chicago faculty since 1900; together, Dickson and Bliss maintained Chicago's position as one of our three leading mathematical centers long after Moore (who remained chairman until 1931) had ceased to publish; see §§13-14.

6. Princeton: Fine, Eisenhart, and Veblen. During the decade 1900-10, Princeton joined Chicago and Harvard as one of our three leading centers

---

*Maschke's career was written up by Bolza in Bull. AMS 15 (1909) 85-95.

of mathematical research. This change was due to the good taste of Henry Burchard Fine (1858-1928).* Fine's love of mathematics apparently had as its source the enthusiastic lectures of George Bruce Halsted (1853-1922), who taught at Princeton from 1879 to 1884 after studying under Sylvester at JHU. Fine got his Ph.D. at Leipzig in 1884, working with Felix Klein and Study, but it was Kronecker's lectures that Fine, like E. H. Moore a year or two later, found most inspiring. Meanwhile Halsted† went to the University of Texas, where he stayed until being fired in 1904 for hiring R. L. Moore over the opposition of the Board of Regents!

Fine was on the Princeton faculty from 1885 until his death in 1928, acting as dean from 1903 on. Though he wrote a number of excellent textbooks, outstanding in their time, he was more important as an administrator than as a scholar. He was a lifelong friend of Woodrow Wilson, President of Princeton and later President of the United States, and acted as dean from 1903 until his death in 1928. Among other things, he raised $3,000,000 for science at Princeton, and Fine Hall is named for him.

In 1900, even before becoming dean, Fine showed his good taste by inviting Luther Pfahler Eisenhart (1878-1965) to join the Princeton faculty. In 1905, Fine and Eisenhart were joined by Veblen and Bliss, while J. H. M. Wedderburn (1882-1948) came to Princeton from Chicago a few years later. After Bliss returned to Chicago (see §5), G. D. Birkhoff came to Princeton from 1909 to 1912 (see §8). Fine was clearly trying to attract to Princeton the best of Chicago's many outstanding Ph.D.'s.

Eisenhart. From this galaxy of outstanding young mathematicians, it was Eisenhart and Veblen who left the most lasting mark on Princeton, making it a great center of geometry and topology. Eisenhart was a master of differential geometry; like Fine, he wrote several excellent books—only Eisenhart's books were graduate texts and surveys. Thus I learned much of what I know about Lie groups from his *Continuous Groups of Transformations* (1933). He also wrote nearly 100 research papers, but it was probably his advanced expository surveys that were most notable. He was President of the AMS in 1931-32, and Dean of the Princeton Graduate School from 1938 on. His activity never stopped; his last paper (on differential geometry and general relativity) was published in 1963, when he was 84 years old!

Veblen.‡ Although Eisenhart was an important figure throughout his

---

*For biographies of Fine, see [1, pp. 167-70] and O. Veblen, Bull. AMS 35 (1929) 726-30. From 1905 on, Fine relied increasingly on Veblen for advice.

†For more information about Halsted, see [17, pp. 123-9] and R. C. Archibald, Scripta Math. 2 (1934) 369.

‡For more about Veblen's career, see Deane Montgomery, Bull. AMS 69 (1963) 26-36, and Saunders Mac Lane [5, vol. 37 (1964) 325-42].

long life, it was Oswald Veblen whose influence was predominant, not only on Princeton but also on a large sector of American mathematics. A son of a professor of physics at the University of Iowa, he got a second A.B. at Harvard in 1900 after graduating from the University of Iowa two years earlier, just before his 18th birthday! He then went to Chicago, where his uncle Thorstein was a prominent (and liberal) professor of economics.

Here he worked closely with E. H. Moore, under whose guidance he wrote an impressive thesis on axioms for Euclidean geometry. This was a significant extension of Moore's systematic efforts to improve Hilbert's *Grundlagen der Geometrie*;‡ other offshoots were collaborative papers with W. H. Bussey, J. H. Maclagan-Wedderburn and Moore's brother-in-law, J. W. Young (1879–1932). The end-product of this collaborative research was Veblen's classic two volume treatise (1910, 1918) on *Projective Geometry*. The first volume, written in collaboration with J. W. Young, contains a masterful discussion of the foundations of the subject, and its concern with projective "geometries" over arbitrary fields relates it directly to Wedderburn's fundamental research, which I shall discuss shortly.

Another early paper by Veblen* contains the first truly rigorous proof of the Jordan curve theorem, foreshadowing his lifelong interest in combinatorial topology (or *analysis situs,* as it was then called). However, his activity in this area did not truly begin until 1912, when he wrote several papers on the subject, of which one with J. W. Alexander on the classification of manifolds (Annals of Math. 14 (1913) 163–78) was especially seminal.

Wedderburn. While at Chicago (see §5), Wedderburn had already collaborated with Veblen in correcting and improving Hilbert's *Grundlagen der Geometrie* (see Trans. AMS 8 (1907) 379–88). It was not unnatural that Veblen should have encouraged Fine to invite Wedderburn to Princeton. Shortly after, Wedderburn wrote his masterpiece, † in which he showed that the most important structure theorems about linear associative algebras could be proved by rational methods, and hence were valid for linear algebras over any field. Since then, the structure theory of such algebras has been referred to as Wedderburn theory. A few years later, Wedderburn also gave a near-proof of the unique factorization theorem for finite groups, also a classic. However, like Dickson (who worked along similar lines) he had no taste for committee work; he was primarily a scholar who was not

---

‡For a Göttingen appraisal of these American contributions, see A. Schmidt in Hilbert's *Ges. Werke,* vol. 2, pp. 404–14.

*Trans. AMS 6 (1905) 83–98. Cf. R. L. Moore, *ibid*, 16 (1915) 27–32; both men were contemporary students under E. H. Moore.

†Proc. Lond. Math. Soc. 6 (1907) 77–118.

deeply concerned about American mathematical development or American institutions as such.

7. George David Birkhoff. The last of Chicago's outstanding Ph.D.'s in the Moore-Bolza-Maschke era was my father, George David Birkhoff (1884–1944). Like Veblen, he was the son of a professional man; both of their fathers' fathers worked with their hands. Veblen was the eldest of eight; G. D. Birkhoff the eldest of seven children. Both were tall, erect, blue-eyed, blond, sociable extroverts with deep interest in all aspects of mathematics and mathematical physics.

G. D. Birkhoff's father David Birkhoff (1859–1909) came to Chicago from Holland in 1871 with his father, who had decided that the New World offered more opportunities to a skilled carpenter than did his native country, where he was umemployed. David's father arrived just in time to help build the first house completed after the Great Chicago Fire.

David earned his way through Rush Medical College by working in a furniture factory. He then married Jennie Droppers, likewise of Dutch ancestry, who had also wished to become a doctor. Following two years in the Dutch community of Overisel, Michigan, he practiced medicine in Chicago until his untimely death. G. D. Birkhoff was born in Overisel, and heard much Dutch spoken around him in his youth. Many of his father's patients were of Dutch extraction, and in 1893, his grandfather's brothers and sister came to visit their brother, and see the Chicago World's Fair and the New World.

G. D. Birkhoff attended Lewis Institute, a partially endowed private school with high educational ideals. It was there, around the age of fifteen, that he fell in love with mathematics; as is explained by Vandiver [19], he sent to the then new *American Mathematical Monthly* a solution to a difficult problem when only 14 or 15! His intense, hardworking father and his gregarious grandfather were both very proud of his mathematical enthusiasm and originality.

When he entered the University of Chicago two years later, in 1901, he was already trying to solve research problems. He presented his first paper to the American Mathematical Society (on number theory) when only 18 years old; his co-author, H. S. Vandiver, later became a very distinguished number theorist. The results of this paper enabled Wedderburn to give the first proof of the theorem that every finite division ring is commutative (see §5).

In [2, p. 274], G. D. Birkhoff has described the thrill he got from his first sight of the "well-filled shelves" at the University of Chicago, and his inspiring teachers there "under the general leadership of E. H. Moore", who "emphasized the abstract and algebraic side of mathematics, although ... remarkably catholic in his outlook". He transferred to Harvard the next year "with Moore's approval". Whether his transfer was influenced by Veblen, who had graduated from Harvard two years before, I do not know. He may also have been influenced by the example of his distinguished uncle Garrett Droppers, who had graduated from Harvard in 1887 and became an economist.† Or, he may have just wanted to learn from Osgood and Bôcher, the leading American analysts at that time.

In his first year at Harvard, still only 19, G. D. Birkhoff formulated the 'Birkhoff-Hermite problem' of determining which sets of $(r+1)$ pairs (i,j) of nonnegative integers have the property that, for arbitrary real c_{ij} and distinct real a_i, there is one and only one polynomial $p(x)$ of degree r satisfying $p^{(j)}(a_i) = c_{ij}$. He observed that mean value and remainder theorems hold in all such cases. Pólya solved the two-point Birkhoff–Hermite problem in 1937; it has recently become very fashionable, and is still unsolved.

At Harvard, G. D. Birkhoff was most stimulated by Bôcher, whose influence is evident in the first major phase of his work—that concerned with "asymptotic expansions, boundary value problems, and Sturm-Liouville theorems."* He also took at least one course with Osgood, and probably with Byerly as well. He once confided to me that he found Osgood's assignments dull, and didn't do them until Osgood threatened to exclude him from the course if he didn't conform!

Having mastered Harvard's offerings, he graduated in 1905. His Harvard classmates included Walter Sherman Gifford, later president of Am. Tel. and Tel., and Clarence Dillon of Dillon, Read and Co., while James K. Rand, the founder of Remington Rand, was a fraternity mate. However, G. D. Birkhoff was never very impressed by business success, perhaps because it was easier to become rich in his lifetime.‡

G. D. Birkhoff's ambition was for *mathematical* distinction, and he continued his quest for this as a graduate student at the University of Chicago in 1905-7. The extent of his ambition is shown by a statement he made years later to me in an expansive moment: that, when studying for his Ph.D. Orals in 1907, he "had tried to master essentially all the mathematics that was known at the time," and had made a good stab at achieving this ambition!

During these years, E. H. Moore was his Thesis Advisor, and he "saw Moore's program of General Analysis taking shape day by day" [6, p. 275].

---

†Though J. P. Morgan's partner Thomas Lamont was his classmate, Garrett Droppers' political views were somewhat similar to those of Thorstein Veblen. He was an active Democrat, and reputed to have influenced William Jennings Bryan to swing his support to Woodrow Wilson, whose Minister to Greece he later became. I was named for Garrett Droppers.

*I am following Morse's analysis [6, vol. 1, pp. xxvi–xxx] (see the footnotes on pp. xxvii, 18, 37, 51, 79, of the volume cited).

‡Thus his Uncle George rose from immigrant to President of the Chicago Board of Real Estate; another uncle (by marriage) co-founded the Scully Iron and Steel Co.; and George Birkhoff's daughter Genevieve married J. Motley Morehead, a poor boy who became a major executive in Union Carbide and benefactor of the University of North Carolina.

However, he was not attracted by the generality and abstractness of Moore's ideas; his Thesis owes much more to Bôcher. Morever, it does not seem that E. H. Moore recognized his preeminence. When E. B. van Vleck offered him an instructorship at Wisconsin (at a salary of $1000), van Vleck was not following E. H. Moore's recommendation. By this time he was engaged to Marjorie Grafius, an alumna of Lewis Institute and the University of Illinois. The engaged couple waited a year to get married, so that they could save enough money to furnish their first home in Madison.

In 1909, the young Birkhoffs moved to Princeton, and it was here that G. D. Birkhoff really found himself. His interests were not limited to analysis. Thus he always liked to tackle difficult unsolved problems, and his papers typically attacked these with radically new ideas and methods. He initiated a novel approach to the four-color problem when 27, observing that the number of ways of coloring a map M in λ colors is a polynomial in λ, the 'chromatic polynomial' $p_M(\lambda)$. A year later, he introduced the concept of a *reducible ring* of n regions with reducing number $\phi(n)$, such that the problem of coloring any map containing such a ring having more than $\phi(n)$ regions inside and outside can be reduced to that of coloring a smaller map. This is the key concept of the celebrated Haken-Appel 1976 computer-aided proof of the four-color theorem.†

G. D. Birkhoff sometimes omitted details of proofs, a trait which led J. D. Tamarkin to claim that one of his early proofs was incomplete. He promptly provided the details.‡

Much of G. D. Birkhoff's most famous work concerned the general properties of (conservative) *dynamical systems*. Already in the summer of 1909, he was led by a paper of Hadamard to introduce the notions of α- and ω-limit points. During the next three years, while at Princeton with Veblen, he continued to think about related questions among many others; the frequent discussions of the two friends ranged over the whole of mathematics!

Finally in 1912, he achieved a most dramatic success by proving Poincaré's conjecture: that any area-preserving transformation of an annulus into itself, which moved the two bounding circles in opposite directions, had to have a fixed point. Since Poincaré and other leading European mathematicians had tried in vain to prove this conjecture, its demonstration by a 28 year old American who had never even studied in Europe made him internationally famous overnight.* Moreover, since Poincaré's conjecture has a direct

---

†Am. J. Math. 35 (1913) 115-28, or [5, vol. iii, 6-19]. He evidently discussed the problem with his friend Veblen; cf. O. Veblen, Annals of Math. 14 (1912) 86-94.

‡See [6, vol. i, pp. xxviii and 78-89]; G. D. Birkhoff had stated frankly (ibid., p. 20, footnote) that "In this proof, certain details of logic are slurred over." My mother told me that G. D. Birkhoff composed his rebuttal while troubled with a painful sty in his eye, perhaps stimulated by emotional stress!

*I am indebted to Richard Courant for his recollection of this event; cf. [30].

application to celestial mechanics (by Liouville's theorem, any conservative Lagrangian system induces a volume-conserving flow on its phase-space), this proof made him the acknowledged heir of the great tradition of Newton, Laplace, Gauss, Hamilton, and Jacobi.

In the same year, G. D. Birkhoff moved from Princeton to Harvard, and the formative stage of his career was over.

YEARS OF TRANSITION: 1914–1920

8. Growing Pains. By 1914, our country's higher mathematics was progressing rapidly. Nearly 25 new Ph.D.'s were being produced annually, mostly in a handful of major centers. This rate of production would continue with slight change until 1925. In an invaluable contemporary appraisal [9a], a committee with Bôcher as chairman assessed the quality as fully up to European standards. However, it also stressed the value of European study for those wishing a broader and deeper mathematical culture.

Alas, after a century of relative peace, during which European science and industry had grown to unprecedented levels, came World War I. It exacted a tragic toll on every phase of life; a whole generation was sacrificed in a futile struggle, leaving deep and permanent scars.

Our country's rapid mathematical progress, on the contrary, continued during the years 1914–1920 with only a brief pause. One result was the end of the tradition of going to Germany for a Ph.D.; by 1920, we had several good graduate centers of our own!

However, our progress was accompanied by some growning pains: with growth came factionalism. The leadership of the AMS was divided on two issues: regional balance, and concern with mathematical education. In particular, the Chicago Section felt neglected by the AMS. Although it had produced the lion's share of outstanding American-educated Ph.D.'s, and the center of population was steadily moving westward, meetings were usually held to suit the convenience of the eastern Ivy League faculties, rather than that of Chicago, Northwestern, and the "big ten" middle western state universities.

E. B. van Vleck. Fortunately, a split along regional lines was avoided, thanks to E. B. van Vleck, AMS President in 1913 and 1914. The tactful way in which he prevented Chicago's sense of neglect from causing a schism is described in [1, pp. 78-81, p. 109]. To this description, I should like to add a few paragraphs about his career and the van Vleck family.

E. B. van Vleck was the second of three generations of distinguished American scientists. His father John M. van Vleck (1833–1912) was, like his contemporaries Hill and Newcomb, an astronomer who had worked for the Nautical Almanac Office in Cambridge. He was for many years a dynamic administrator at Wesleyan, and later active in the AMS, of which he was vice-president in 1904. John M. van Vleck's students included Henry S. White* (1861–1943) and F. S. Woods, who together with E. B. van Vleck gave the AMS colloquium in 1903. White became the ninth AMS president four years later; he was still faithfully attending AMS meetings in the mid-1930's when over 75, chatting benevolently with aspiring young mathematicians like myself!

After graduating from Wesleyan in 1884, E. B. van Vleck (1863–1943) stayed on a year as an assistant in physics. He next spent two years at JHU, studying and serving as an assistant to the experimental physicist Rowland, and then three years at Wesleyan as a teaching assistant, before finally going to Göttingen in 1890-93, where he studied "with Burkhardt, Fricke, F. Schur, H. A. Schwarz, Voigt, Weber, and that 'marvelous teacher Felix Klein'" [1, p. 171], finally getting his Ph.D. at age 30.** He taught mathematics at the University of Wisconsin from 1893 to 1929; his mathematical publications, though not voluminous, were skillfully written and widely read. He may have anticipated Borel by proving the zero-one law.‡ I remember him well as a kindly man, who gave my father his first job. † He was elected to the National Academy in 1911, and became vice-president of the AAAS. He also travelled all over the world and acquired an outstanding collection of Japanese prints.

E. B. van Vleck's son John H. van Vleck (1899–) is a world-famous expert on magnetism. He has been president of the American Physical Society and received many other honors.

E. W. Brown. The next AMS president was Ernest William Brown (1866–1938), who had come to Haverford College from England in 1891, following Frank Morley (1860-1937). Morley moved to Johns Hopkins in 1900, and Brown to Yale in 1907. Presumably, Brown was chosen as AMS president in preference to his colleague James Pierpont, who had been a major factor in founding the *Transactions,* because of his greater research reputation. Morley would become AMS president in 1919-20.

Brown was the third and last astronomer to be AMS president (Fine had been acting director of the Princeton Observatory in 1908-12, also serving as AMS president in 1911-12.) Like his predecessor G. W. Hill, Brown was a lifelong bachelor; for comments on this fact (written by another bachelor) see [1, p. 177]; perhaps evenings spent in lengthy calculations are not conducive

---

*See Bull. AMS 49 (1943) 670-1.

**For his biography, see [21, vol. 30, pp. 399–409] also G. D. Birkhoff, Bull. AMS 50 (1944) 37-41.

‡See A. Novikoff and J. Barone, Historia Math., 4 (1977) 43-65.

†Legend has it that E. H. Moore recommended someone else, but that E. B. van Vleck said 'I'll take the Dutchman'. This sounds apocryphal!

to matrimony! "During a year of post-graduate work at Cambridge [University], his chief adviser, Prof. G. H. Darwin, recommended him to study G. W. Hill's classic paper...and thus he started in a field of research which was to occupy him for more than forty-five years." [1, p. 174]. Namely, he carried on Hill's tradition of making improved predictions of the orbits of the planets and their moons. He won the Gold Medal of the Royal Astronomical Society in 1907, for developing new algorithms for calculating the position of the moon. This led to an invitation to leave Haverford for Yale, where he taught for 25 years, and where his distinguished tradition was continued until recently by Dirk Brouwer.*

After 1920, celestial mechanics ceased to dominate theoretical astronomy, which thereby lost its traditional affiliation with mathematics. Thus when Thomas J. Watson established a center for astronomical computations at Columbia in 1929-33 [12, p. 109] under W. J. Eckert (1902-71), the event went almost unnoticed in mathematical circles.

Indeed, recent Ph.D.'s in pure mathematics may wonder why celestial mechanics was esteemed so highly before 1920! They should remember how such precise calculations led not only to the discovery of Neptune, but also to that of the advance in the perihelion of Mercury, one of the main inspirations for Einstein's general theory of relativity. Celestial mechanics was the first branch of mathematical physics to be made exact—and until 1900, mathematical physics was considered to be a branch of mathematics. Even in our time, the design of accelerators and the control of satellites depends on methods similar to those developed by G. W. Hill, Simon Newcomb, and E. W. Brown.‡

9. The Monthly and the MAA.† In 1893, Benjamin Finkel and J. M. Colaw, both members of the NYMS, published the first issue of the *American Mathematical Monthly* in Kidder, Missouri. It was "devoted to the solution of problems in pure and applied mathematics, papers on mathematical subjects, biographies of noted mathematicians, etc." From the start, it stimulated the curiosity and sharpened the wits of many aspiring young mathematicians, including G. D. Birkhoff and H. S. Vandiver.

During the first decades of the AMS, many of its leaders were active in mathematical education. Thus David Eugene Smith of Columbia, AMS Librarian 1902-1920 and Vice President in 1922, was an eminently successful

---

*See the articles by Eckert (Brown's student) and by Brouwer and Clemence in Proc. IX Symp. Applied Math., Amer. Math. Soc., 1959.

‡Ibid; see especially the articles by E. D. Courant, F. W. Shipple, K. A. Ericke, and J. W. Siry.

†For a fuller account, see [7, Chap. I]. Also Amer. Math. Monthly 64 (1933) No. 7: "The Otto Dunkel Memorial Problem Book," (H. Eves and E. P. Starke, eds.). Otto Dunkel was another Bôcher Ph.D.

writer of high school textbooks. Fiske was secretary of the College Entrance Examination Board from 1902 to 1936. Moreover E. H. Moore had some original ideas about the teaching of elementary mathematics. Thus his retiring AMS presidential address (1902) dealt with high-school and junior college education.‡ Though given a chilly reception by its auditors [14, p. 16], its ideas were widely disseminated by Herbert Ellsworth Slaught, who was on the Chicago faculty from 1898 to 1931, and thus had considerable national influence.

In 1903, L. E. Dickson joined the editorial staff of the *Monthly*, and in 1906 the University of Chicago assumed exclusive control of it, with Slaught as Chief Editor from 1908 on. Thus in 1910, Chicago ran the *Monthly*, Harvard ran the *Annals*, JHU the *American Journal of Mathematics*, and the AMS ran the *Transactions*.

Slaught. It was above all H. E. Slaught (1861–1937) who was active in organizing mathematical education in our country.† He began by inviting the University of Illinois to co-sponsor the *Monthly*. In 1914, he invited the AMS to broaden its base while strengthening the *Monthly* by sponsoring this journal. This effort was unsuccessful, partly because of the strong opposition of Osgood, at the time a dominant figure of the Eastern establishment [1, p. 79]. From then on the AMS officially dissociated itself from the problems of high-school and undergraduate college education.

The Chicago Section was already resentful at the tendency of Eastern mathematicians to ignore the rest of our country (see §8). Following the refusal of the AMS to sponsor the *Monthly*, Slaught led a movement that resulted in the foundation of the *Mathematical Association of America* (MAA).

Slaught's activity did not stop there; in 1920, he organized the National Council of Teachers of Mathematics (NCTM); and a few years later, the Carus Monograph Series. Small wonder that, before he died, he was made permanent honorary president of both the MAA and the NCTM!

Earle Raymond Hedrick. The first President of the MAA was Earle Raymond Hedrick (1876–1943), a many-sided man with varied accomplishments. After studying at the University of Michigan and Harvard, he got a Ph.D. at Göttingen, after which he spent some months in Paris, "in contact with such men as Goursat, Picard, Hadamard, Appell, and Jules Tannery" [1, p. 223]. One of his first contributions was to translate Goursat's *Cours d'*

---

‡Reprinted in *Readings in the History of Mathematics Education*, J. K. Bidwell and R. G. Clason (eds), National Council of Teachers of Mathematics, 1970.

†For tributes to his work, see Amer. Math. Monthly 45 (1938) 1–10; also G. A. Bliss in Bull. AMS 43 (1937) 595–7. He was at the University of Chicago from 1892 on!

Analyse into English, thus making this classic treatise more accessible to American students. While continuing with research, though gradually less and less, he became increasingly influential by popularizing advanced mathematical ideas through his articles and textbooks, and by his organizational activities. He was editor of the *Bulletin* from 1921 to 1937, had nine children, drank legendary quantities of coffee, and left the University of Missouri to become Chairman of the Mathematics Department at the (then new) University of California at Los Angeles in 1924. When he became Provost of that institution, he finally gave up being editor of the *Bulletin*. He was President of the AMS in 1929–30, and Vice-President of AAAS in 1931. In all these roles, he was a dynamic man who "got things done" in a true pioneer spirit.

Hedrick was outstanding in his leadership, but many others were instrumental in spreading the mathematical gospel to the far reaches of the country (as seen from the northeast). Among these Griffith C. Evans (1887–1973) and Lester R. Ford (1886–1967) were especially notable. Both Ph.D.'s of Bôcher, † they went to Rice University in Houston and married two sisters, granddaughters of Sam Houston. Ford later became President of the MAA and Evans President of the AMS. Ford's *Automorphic Functions*, published in 1929 as an elaboration of a 1915 tract, is still considered outstanding today (see Bull. AMS 82 (1976) 218). Evans published *two* volumes in the AMS colloquium series: *Functionals and their Applications* (1918), and *The Logarithmic Potential* (1928). He moved to Berkeley as Chairman in 1934, and built it up into one of our "big four" mathematical centers.

However, all this happened long after 1914, and I want now to recall four leaders of a very different stamp, whose constructive influence on American mathematics was strong during the years 1913–41.

10. Topology takes root. During the decade 1910–1920, topology became the primary concern of several leading American mathematicians. Veblen, R. L. Moore, Alexander, and Lefschetz were especially influential converts to the subject, as I shall now try to explain. During the same years that G. D. Birkhoff shifted his center of interest to dynamical systems, Veblen changed his from projective geometry to topology. Thus in 1912–13, he established with J. W. Alexander a seminal duality theorem, whose proof used the (then novel) technique of considering homology groups mod 2.* In 1916, he gave the colloquium lectures on *Analysis Situs,* but it was not until 1922 that these appeared in book form.

In the meantime, he had finished the second volume of his *Projective Geometry,* a great classic whose preface still echoes Klein's *Erlangen Programm* and Hilbert's *Grundlagen der Geometrie.*

---

†[1, p. 163.] Actually, Ford completed his Thesis in Paris under Humbert.
*It had been invented a few years before by H. Tietze, Monats. Math. Phys. 19 (1908) p. 49.

He had also been a leader in our mathematical World War I effort, having been in charge (with F. R. Moulton) of range firing and ballistic work at the Aberdeen Proving Ground (after giving his AMS Colloquium Lectures on topology in 1916). His staff included Alexander, Bennett, Blichfeldt, Bliss, Dines, Franklin, Graustein, Gronwall, Hart, Haskins, Jackson, Milne, Mitchell, Ritt, Roever, and Vandiver,* all noteworthy mathematicians; H. H. Goldstine [12, Ch. 9] has given an excellent first-hand account of their roles. Both Veblen and Bliss published papers based on their contributions to this wartime work; even more notable was the contribution of F. R. Moulton, the Chicago mathematical astronomer. ‡ In Ch. III of his *New Methods of Exterior Ballistics* (Univ. of Chicago Press, 1926), he explained the so-called Adams-Moulton method, still the best multistep method for numerically integrating systems of ordinary differential equations.

R. L. Moore. A second American topologist of note was R. L. Moore (1882-1974). Texan to the core, Moore was a rugged individualist whose 1905 Thesis,† written under Veblen, was an extension of Veblen's 1904 Thesis on the foundations of Euclidean geometry, written under E. H. Moore. In 1915, R. L. Moore made his debut as a topologist by extending Veblen's discussion of the Jordan curve theorem. This set the tone of his later research on the foundations of point-set topology, in which he treated the plane with especial affection.** His methods were highly original, and his results often anticipated (at least in special cases), theorems commonly attributed to European mathematicians (e.g., Zorn's Lemma and Urysohn's metrization theorem). He encouraged originality in his students by forbidding them to read standard expositions, and requiring them to "work things out" by themselves in a true pioneer spirit instead. This unconventional procedure was highly successful in producing research mathematicians; Moore's Ph.D.'s included such important topologists as R. L. Wilder, Gordon T. Whyburn, and J. R. Kline (for many years AMS Secretary), Gail Young, R. H. Bing, E. E. Moise, Eldon Dyer and Mary Rudin.

J. W. Alexander. Following his seminal 1913 paper with Veblen, Alexander published two or three others on combinatorial topology before writing a

---

*See [1, p. 209], and L. E. Dickson, Bull. AMS 25 (1919) 289-311, esp. p. 296.

‡Moulton had stimulated Bliss' first published paper, and been AMS vice-president in 1915.

†Trans. AMS 9 (1908) 487-512. G. D. Birkhoff's *Basic Geometry* and the author's "Metric foundations of geometry" (Trans. AMS 55 (1944) 465-92) are in the same tradition. For his treatment of the Jordan curve theorem see *ibid.* 16 (1915) 27-32.

**His "Foundations of Point-Set Theory," Amer. Math. Soc., 1932, and Whyburn's "Analytic Topology," Amer. Math. Soc., 1942, give a good idea of Moore's interests. For a biographical sketch, see R. L. Wilder, Bull. AMS 82 (1976) 417-27.

Ph.D. thesis which was, curiously, on complex analytic functions† and written under Gronwall, who had replaced G. D. Birkhoff at Princeton. Alexander also made most of the drawings for Vol. II of Veblen's *Projective Geometry*, and served under him at Aberdeen.

In the 1920's, Alexander became an expert on knot theory, and extended his earlier duality theorem (with Veblen) to the n-dimensional generalization by Brouwer of the Jordan curve theorem, the so-called Jordan-Brouwer Theorem (see Math. Annalen 71 (1912) 314-27; for Alexander's contributions see Trans. AMS 23 (1922) 333-49).

Alexander continued to make original contributions to topology until World War II. Thus Pontrjagin states that "the initiative for using continuous groups in combinatorial topology belongs to J. W. Alexander and L. W. Cohen." It is for me a special personal pleasure to pay tribute to this charming and brilliant man, a millionaire with liberal political views and a rare collection of limericks, who was a frequent visitor at the von Neumann house when I was a guest there in the 1930's. Though he was nearly 25 years older, his spirit was so young that I thought of him as a near-contemporary!

Lefschetz. Equally original, and far more influential on the international scene was Solomon Lefschetz (1884-1972). Born in Russia and educated in Paris, Lefschetz worked in industry for several years, losing both hands in an industrial accident before he turned to mathematics. He was 29 before his thesis "On the existence of loci with given singularities" was published (Trans. AMS 14 (1913) 23-41). He became famous six years later, when he won a prize in Paris for showing that algebraic homology underlay a relation of equivalence introduced into algebraic geometry by Severi, its acknowledged master. In Lefschetz' own colorful language, he "planted the harpoon of topology into the belly of the whale of algebraic geometry."‡ His prize paper was translated into English and republished, with minor modifications, in Trans. AMS 22 (1921) 326-482; its final form was a Borel monograph *L'Analysis Situs et la Géometrie Algébrique*, Paris, 1924.

In 1924, he joined Veblen and Alexander at Princeton, making it the world's leading center of topology. His Colloquium volume on "Topology" (1930) was an early fruit of this phase of his life; Paul Smith, A. W. Tucker, and R. J. Walker were even more important products. In the 1930's, Lefschetz played an even more active role as an *international* leader, inspiring Norman

---

†See Annals of Math. 17 (1915) 12-22.

‡S. Lefschetz, Bull. AMS 74 (1968) 854-79. Other reminiscences are in Amer. Math. Monthly, 77 (1970) 344-50. Biographical sketches of Lefschetz have been written by L. Markus (Bull. AMS 79 (1973) 663- and W. V. D. Hodge, Roy. Soc. (1973). See also "Algebraic Geometry and Topology," R. H. Fox, D. C. Spencer, and A. W. Tucker (eds.), Princeton University Press, 1957, for other appreciations of Lefschetz' work.

Steenrod and Henry Wallman in the process—but I am getting ahead of myself (see §20).

11. Cambridge: 1912–1941. During the years 1906–1941, a large fraction of our leading research mathematicians lived near either Chicago, where the University of Chicago and Northwestern are located; Princeton, where the Institute for Advanced Study became a neighbor of Princeton University in the 1930's; or Cambridge, the home of Harvard and M.I.T. I shall next record some impressions of faculty life in Cambridge,† where G. D. Birkhoff joined Bôcher and Osgood in 1912. I imagine that life at Princeton and Northwestern followed much the same pattern.

In retrospect, it seems unbelievably tranquil. The Harvard faculty still formed a cohesive group; most of its members lived within walking distance of Harvard and each other. Department meetings were held at the home of the Chairman, and most professors lunched at home; afternoon naps were not uncommon. Although most professors taught nine hours a week until 1926, and probably worked much harder than their European colleagues,‡ there were few distracting administrative chores; in 1930, one efficient half-time secretary was still able to type all the letters and research papers of the Mathematics Department.

The open country was close by. Thus the banks of the Charles, which had been tidal flats until a few years before, were still in a state of nature, and the marshy outskirts of West Cambridge were not yet built up. Milk came in by horse and wagon from dairies in nearby Lexington and Lincoln, or by train from New Hampshire or Vermont. The pace of life was leisurely: a three months summer vacation at the seashore or among New England mountains and lakes was part of the normal rhythm.

In this tranquil environment, G. D. Birkhoff developed his ideas about differential equations and dynamical systems from 1912 to 1920, occasionally taking time out to go to scientific meetings or to think about the four color problem. During these years Morse and Walsh came as students; Bôcher and Byerly died, and W. C. Graustein joined the faculty.

Around 1920, G. D. Birkhoff was promoted to a full professorship, and moved from a row house across from the Radcliffe dormitories to a 12 room house a stone's throw from stately Brattle St. Most full professors of that era had live-in maids and entertained at Victorian dinner parties; the Birkhoffs were no exception. Moreover the 250 or so tenured ('permanent') faculty members knew each other, and faculty meetings were almost family gatherings, enlivened by the wit of familiar colorful personalities. There was a Shop Club, at which professors from the Law and Business Schools ex-

---

†Wiener [22] gives earlier impressions of the Cambridge environment as he saw it.
‡See p. 51, footnote ‡.

changed informal accounts of their activities and interests with their colleagues from the Faculty of Arts and Sciences.

Until World War II, the Harvard Mathematics Department was dominated by Harvard graduates such as Osgood, Coolidge, Huntington, G. D. Birkhoff, Walsh, Graustein, and Stone, all of whom were on the faculty in the 1930's. This made faculty interest in undergraduate teaching and advising very natural, and gave a strong sense of stability and continuity. Moreover research was done quietly at home. Whereas today at least ten seminars in Cambridge compete for attention, each concerned with a different subarea of mathematical research, there was then only one weekly colloquium in Greater Boston. It was attended also by research-oriented M.I.T. staff members and (often) by a contingent from Brown.

M.I.T. The atmosphere of M.I.T. was very different. Located in industrial East Cambridge, its faculty commuted from Belmont or Waverly. Industrial problems and opportunities were its life's blood, and some of its graduates and staff built up substantial businesses in the Boston area, like the Dewey and Almy Chemical Co. and Arthur D. Little.

One never felt isolated at either Harvard or M.I.T. Both universities were within minutes of downtown Boston by subway. For a 20¢ round trip fare, one could get in 20 minutes to any of five or more legitimate theaters, as an alternative to a bird-watching walk. For $3.50, one could take the Fall River night boat to New York, arriving in good time for the AMS meetings unless there happened to be a bad fog.

12. Four cousins. Harvard's great mathematical strength during the years 1912–41 owed much to the intellectual, moral and financial support of President Lowell and three of his cousins. As a substantial beneficiary of this support, I take special pleasure in recalling their personalities and influence.

Abbott Lawrence Lowell. You will recall that Charles W. Eliot, Harvard's president from 1869 to 1908, was a chemist who had worked in the Nautical Almanac Office. His successor Abbott Lawrence Lowell (1860-1945), Harvard's president from 1908 to 1933, was much more mathematically minded. A well-born Bostonian who majored in mathematics at Harvard, he was deeply impressed by Benjamin Peirce, "the most profoundly inspiring teacher I ever had" [5, p. 271]. He graduated *summa cum laude,* and his thesis was published in the *Proceedings* of the American Academy of Arts and Sciences in Boston (vol. 13 (1877) pp. 222-50), eight years before E. H. Moore's Ph.D. Thesis was published in the *Transactions* of the Connecticut Academy of Arts and Sciences.

Lowell went on into law and government; his book about British government and Lord Bryce's book on American government were considered,

for many years, to be the best books on their respective subjects. His constructive influence on mathematics began shortly after he succeeded Eliot. He appointed a committee, chaired by his sister's husband W. L. Putnam, to reform the Harvard mathematics curriculum. Following a broad hint from Lowell, this committee recommended that the normal freshman course should consist of analytic geometry and the *calculus*, which Lowell considered (along with the phonetic alphabet and the Hindu-Arabic decimal notation for numbers) to be one of the greatest inventions of all time.†
Lowell followed up his initial reform by requiring every Harvard undergraduate to take at least one course in mathematics or philosophy, and by advising class after class of entering freshmen in his welcoming address that there was no better preparation for the law than the study of mathematics.

The boldness of this step can be better appreciated if one realizes that in 1911, any course *above* first-year calculus was considered an acceptable part of graduate study [9a]. Even in the 1930's, mathematics concentrators in many small colleges only got to the calculus in their senior year! But Lowell was never a traditionalist; his collected essays on education were published under the title "At War with Academic Traditions in America."

In 1925, Lowell introduced tutorial instruction, aimed at making Harvard undergraduate education less imitative and more reflective. The Harvard Mathematics Department agreed to participate in the tutorial system. By attributing 3 hours a week to the supervision of graduate students and 1.5 hours to tutorial, it got its teaching load reduced from 9 to 4.5 classroom hours a week. This was almost unique in our country at that time. Thus J. C. Fields of Fields Medal fame had complained in 1919 that the average U. S. or Canadian college professor taught 400 hours a year, as compared with 100 in the more advance European countries.‡

In 1933, Lowell implemented his most valuable idea: the establishment of a Society of Fellows where outstanding men in their twenties would get every encouragement to develop their own ideas.* Whereas the German Ph.D. system had *trained* men to do research, Lowell designed and endowed Harvard's Society of Fellows to *free* outstanding young men to do research. I shall describe its success later (§19).

William Lowell Putnam. The Lowell genes were strong; A. L. Lowell's brother Percival was the discoverer of the planet Pluto,** and his sister Amy

---

†J. W. Gibbs expressed a similar opinion about matrix algebra.

‡J. C. Fields, "Universities, research, and brain waste," (Presidential Address). Trans. Royal Canadian Inst. (1920) 3-27.

*Ulam has described its stimulating atmosphere in [29, Ch. 5].

**See W. G. Hoyt, "Lowell and Mars," Univ. of Arizona Press, 1976.

was a well-known poetess. But most memorable for mathematicians was his brother-in-law and cousin William Lowell Putnam (1861-1924), who graduated from Harvard with high honors in mathematics, and was for years the devoted and hospitable chairman of the Visiting Committee of Harvard's Mathematics Department.

In partial fulfillment of W. L. Putman's expressed wish to make excellence in scholarship as honorable among undergraduates as athletic prowess, his widow and her sons cooperated with G. D. Birkhoff in setting up the Putnam Competition in essentially its present form in 1938. I will not repeat here my attempt [8] at reconstructing the background of this Competition, but will add that W. L. Putnam's tradition of hospitality has been maintained to this day by the George Putnams, his son and grandson.*

J. L. Coolidge. Among A. L. Lowell's cousins was also a prominent mathematician, Julian Lowell Coolidge (1873-1954), who was incidentally a descendant of Thomas Jefferson. After teaching Franklin D. Roosevelt and others at Groton, and serving as a Rough Rider under Theodore Roosevelt in the war of 1898, Coolidge became a Harvard instructor in 1900. He studied geometry abroad with Kowalewski, Study, and Corrado Segre in 1902-4, then returning to Harvard where he became an assistant professor in 1908, at age 35.

Since D. J. Struik has ably summarized elsewhere Coolidge's subsequent mathematical career,† I shall only recall that he was President of the MAA and founder of the Chauvenet Prize. He was also vice-president in 1918, and chief fund-raiser for the AMS in the 1920's (see §13). His many books (published by the Oxford University Press) were lively and widely read, his *Probability* being translated into German.

After he became the first Master of Lowell House‡ in 1930, he devoted his main energies to the human aspects of undergraduate education; I was his Senior Tutor there during 1936-38. However, he continued to teach mathematics, as well as writing books about the history of mathematics. His *History of Geometrical Methods* (1940) is one of a very few dealing on a technical level with mathematical developments during the 19th and 20th centuries.

G. E. Roosevelt. The last of the gifted and public-spirited cousins who

---

*Mrs. W. L. Putnam showed the same spirit of friendly hospitality by giving the G. D. Birkhoffs the use of her summer home at Manchester, Mass., in the summer of 1927.

†Amer. Math. Monthly 62 (1955) 669-82.

‡This was one of the two completely new dormitories built when Harvard's House Plan came into being. The money was provided by Edward Harkness, but I have been told that Lowell had to advance some cash to avoid liquidating oil stocks at a very low price after the stock market crash.

actively fostered mathematics (especially Harvard mathematics) from 1910 on was George Emlen Roosevelt. His father had asked him to join the family banking firm when he graduated from St. Marks, but G. E. Roosevelt secured permission to go to Harvard *if* he stayed on the honor roll. After graduating *summa cum laude* in mathematics, he served under W. L. Putnam on the committee that made calculus standard fare for freshmen. From that time on, he was a dynamic and generous member of our Visiting Committee; I remember his lecturing on navigation to our Math. Club in the best tradition of Nathaniel Bowditch—he was Commodore of the New York Yacht Club, and often raced to Bermuda.

His many activities included managing of the estate of Sara Delano Roosevelt (about which he had an amusing interchange of letters with her son F. D. R. when the latter abandoned the gold standard). He was also active in the affairs of N.Y.U.; Richard Courant (see §21) became his good friend, and he was acting president of NYU during the late 1950's.

I had the privilege of conversing many times with Coolidge, Lowell, George Putnam, Sr., and G. E. Roosevelt. I hope that you will read enough between the lines of what I have written to appreciate their spirit, and how much our country owes to men like them. Though it is dangerous to impute motives to other people, I think it safe to say that it was love of and pride in their country and Harvard, a general spirit of philanthropy, and a personal appreciation of the orderliness and beauty of mathematics, that impelled these cousins to promote American mathematics so fruitfully during their lifetimes.

13. Dawn of an era. By 1920, the war was over and a new era was dawning. In 1903, U.S. mathematicians had ranked E. H. Moore, Hill, Osgood, Bôcher, Bolza, Newcomb, Morley, E. W. Brown, H. S. White, and Dickson (in that order) as best. Of these, the geometer Frank Morley (1860-1937) was AMS president; all the others except Bolza had already served, and a new generation was ready to take over.

Across the Atlantic, they saw a changed Europe. Felix Klein, Hilbert, Weyl, and other mathematicians from the defeated Central Powers were excluded by the victorious Allies from participating in the International Mathematical Congresses of 1920 (Strasbourg) and 1924 (Toronto). Ph.D. training in Germany was no longer attractive, let alone a "must".

The most prominent mathematicians of the new generation were Dickson, Bliss, Veblen, and G. D. Birkhoff, all Chicago Ph.D.'s. Of these, Dickson had preceded Morley as AMS president. His prodigious output now exceeded 200 research papers; in 1914 alone, he had published his 1913 colloquium lectures on *Invariants and the Theory of Numbers* and two other monographs. But most notable among his pre-1920 contributions to number theory was his 1600 page *History of the Theory of Numbers* (3 vols.), published in 1919.

His research in the 1920's centered around number theory in linear associative algebras (as defined by Benjamin Peirce). He presented his ideas in his book on *Algebras and Their Arithmetics* (1923) and its revision in German translation (1927). These were the best books on the subject at the time; see also Proc. Int. Math. Congress Toronto (1924), vol. i, pp. 95–102. Likewise, Dickson's *First Course in the Theory of Equations* (1922) introduced a generation of American students to this subject.

Although Dickson served as President of the American Mathematical Society during the war years of 1917-18, he was never an "organization man," preferring bridge, tennis and billiards to tedious committee meetings. In a similar spirit, he resigned from the American Philosophical Society and National Academy of Sciences, retaining membership only in the American Academy among our three major honorary scientific societies. He was much less active in AMS affairs than Bliss, Veblen, or G. D. Birkhoff, concentrating increasingly on his own research and that of his students. From 1921 to 1937 he wrote eight books and 80 research papers; his Ph.D.'s (some earlier) included C. C. MacDuffee, A. A. Albert, Gordon Pall, Mina Rees, R. D. James, Ralph Hull, and 58 others.

A DECADE OF EXPANSION

14. The new establishment. In 1921, the AMS entered a new phase, with the resignation of F. N. Cole as AMS Secretary and Editor of the *Bulletin*, of which volume 21 was dedicated to him.

R. G. D. Richardson. Hedrick took over Cole's editorial duties, and R. G. D. Richardson (1878–1949)* replaced him as AMS Secretary. A distant relative of Simon Newcomb, he taught in Nova Scotia high schools for some years before going to Yale, where he got his Ph.D. in 1906 with Pierpont. After visiting Göttingen in 1908-9, he wrote a number of papers on differential equations, among which most interesting in retrospect is his "A new method in boundary problems for differential equations."† This foreshadowed the later difinitive paper by Courant, Friedrichs and Lewy, by using difference approximations to establish the existence of solutions of elliptic and hyperbolic problems.

Like Fine, Hedrick and many other American mathematicians of his and earlier generations, Richardson was more notable as a builder of institutions than as an individual research mathematician; for over twenty years, he gave unstintingly of his energy and wisdom, cooperating with Bliss, Veblen, G. D. Birkhoff and others in strengthening the AMS.

---

*See Bull. AMS 56 (1950) 256–65.
†Trans. AMS 18 (1917) 489–508.

The triumvirate. Much as E. H. Moore, Osgood, Bôcher, and Fine were the leading figures in American mathematics from 1900 to 1914, so Veblen, G. D. Birkhoff, and Bliss were its principal leaders from 1921 to 1941. All native midwesterners and Chicago Ph.D.'s, they represented our three most important research centers during this period: Princeton, Harvard, and Chicago. They inaugurated their period of leadership by serving as Presidents of the American Mathematical Society for three successive terms, from 1921 through 1927. Bliss (1876-1951) was four years older than Veblen, and Veblen (1880-1960) four years older than Birkhoff, and they served in the order of their seniority.

The first eight years under the new establishment constituted a remarkable period of solid expansion for the Society: Its membership increased from 770 in 1920 to 1758 in 1928, and its budget from $8865 to $28,400. Together with Veblen, Bliss provided much of the initiative for this great progress. Since it has been eloquently described by Archibald in [**1**, pp. 29-31]. I shall not discuss it further here.

Instead, I want to describe briefly how these men, by this time in their forties, exemplified the ideals of the AMS in their own research during these years.

Bliss and Veblen. In 1921, Bliss had just finished writing up the adjoint method for correcting trajectories he had invented while at Aberdeen (see §10),† and was polishing his notes on the calculus of variations. Some of these were to be published as the first in the series of Carus monographs (see §9), of which Slaught had made Bliss chief editor, and which contributed notably to the education of American graduate students.

In addition, he wrote some 20 papers and several books, some expository, during the years from 1920 to 1936 (by which time he was sixty). These included his AMS Colloquium Lectures on *Algebraic Functions* (1933) and a few related papers, but the calculus of variations (especially the Problem of Bolza) continued to occupy the center of his thoughts. His final *Lectures on the Calculus of Variations*, presenting in book form the final distillation of many sets of mimeographed lecture notes, would not be published until 1946. In the meantime, he had supervised the theses of Graves, Duren‡ McShane, M. S. Hestenes, A. S. Householder, and 41 other Ph.D.'s.

Like Bliss, Veblen had been distracted by his war work at Aberdeen, and he devoted much time to his AMS presidential duties. Perhaps for this reason, he increasingly let others share in developing his ideas. More-

---

†Jour. U. S. Artillery 51 (1919) 296-311 and 445-91; Trans. AMS 21 (1920) 93-106.
‡For Duren's impressions as a graduate student in the 1920's, see Amer. Math. Monthly 83 (1976) 243-8.

over from topology, he moved on to differential geometry during the mid-1920's, perhaps attracted (like many of his contemporaries) by the idea of "geometrizing physics" by a suitable extension of Einstein's general relativity. His most notable students during this period were T. Y. Thomas, the logician Alonzo Church, and J. H. C. Whitehead, with whom he wrote Cambridge Tract #29, *On the Foundations of Differential Geometry.* This contains in Chapter VI a set of axioms for global differential geometry, vaguely foreshadowing modern manifold theory.

G. D. Birkhoff. But it was above all G. D. Birkhoff who exemplified the research ideals of the AMS during the 1920's. For a decade after his great triumph in proving Poincaré's geometric theorem, he was mainly advancing the general theory of dynamical systems with new methods. Notable among these was his "minimax method," whose extensions by his student Marston Morse I shall discuss shortly. For this work, Birkhoff was awarded the first Bôcher Prize, after giving the AMS Colloquium Lectures in 1920. An expanded and matured version of these is contained in his *Dynamical Systems*, published in 1927. Here he uses compactness arguments involving α- and ω- limit points to establish the existence of almost periodic "central motions" in *any* dynamical system. In 1928, he and Paul Smith pointed out the special properties that measure-preserving transformations had when they were "metrically transitive," a key concept of ergodic theory often referred to as "ergodicity."

During these years, his scientific interests extended to other areas. Thus he proved with O. D. Kellogg (who lived next door)* the first fixed point theorem in function space. This foreshadowed the Leray-Schauder theory that was to come a decade later. In the direction of physics, he wrote two original books on Einstein's then revolutionary theories of relativity. He also invented a "perfect fluid", from which he deduced "a formula of the Balmer type for the frequencies" of natural oscillation of a hydrogen atom. For this work, published in 1927, he received the 1926 AAAS prize.

In 1926, the Birkhoffs also made their first trip to Europe, staying there nearly eight months on sabbatical leave. One of G. D. Birkhoff's missions was to survey European mathematical centers with special reference to strengthening their ties with the physical sciences. On the basis of his recommendations the International Board of the Rockefeller Foundation made grants of $275,000 to Göttingen and $322,250 to the University of Paris,† thus partially repaying a long-standing cultural debt.

This 1926 trip inaugurated a long period of international activity during

---

*For Kellogg's work, see G. D. Birkhoff, Bull. AMS 39 (1933) 171-7; [**5**, iii, 537-43].

†See Chapter 12 of Raymond B. Fosdick, *The Story of the Rockefeller Foundation.* Harper, 1952, and W. Weaver, "Early support for mathematics from Rockefeller agencies."

which Birkhoff visited and was honored in countless countries. Already in 1928, Harvard financed a trip around the world, during which he studied and collected oriental music to test his original ideas about the quantitative psychology of aesthetics. (For a definitive statement of these ideas, see his book *Aesthetic Measure*, Harvard University Press, 1932.)

The Bologna Congress. The researches of Veblen and Birkhoff were honored by invitations to give hour addresses at the 1928 International Congress.‡ Veblen's talk summarized his views about differential geometry, emphasizing the relevance of the infinitesimal parallelism of Levi-Civita to Einstein's general relativity, and suggesting that Felix Klein's Erlanger Programm had become played out.†

Birkhoff's talk on "Quelques éléments mathématiques de l'art" summarized his ideas about aesthetics. Given at a special session at Florence in the beautiful Palazzo Vecchio, and heralded by two trumpeters in medieval costumes, it involved no mathematics beyond arithmetic! However, Hadamard had already paid tribute to Birkhoff's technical work, presenting the Birkhoff-Kellogg fixed-point theorem in function space as the crowning achievement of functional analysis up to that time. American mathematics had indeed come a long way since 1891.

The leadership and cooperation that I have been describing were to last for at least another decade, as I shall explain later. But the impression to leave with you now is that of a closely knit American Mathematical Establishment during the years 1921–41, headed by four very broad and distinguished men, united by bonds of personal friendship which dated back to the Chicago of E. H. Moore in the early 1900's, and included wives. Thus the Birkhoffs occupied the Veblen summer cottage in Brooklin, Maine, in the summer of 1925. Walking barefoot from this cottage through a woods path, one came in 5 minutes to the cottage of Nobel Prize winner Davisson, whose wife Lottie was Mrs. Veblen's sister. The widows of G. D. Birkhoff, Veblen, and Richardson were still corresponding intimately and affectionately in the 1970's.

The mathematical world which I am trying to describe was truly on a 'human scale'.

15. Morse and Stone. During his years at Harvard, G. D. Birkhoff had a remarkably distinguished list of Ph.D. students, including Joseph Slepian, Marston Morse, H. J. Ettlinger, J. L. Walsh, Rudolph Langer, D. V. Widder, B. O. Koopman, Marshall Stone, C. B. Morrey, and Hassler

---

‡Birkhoff and Veblen were again to be invited American speakers at the Oslo Congress eight years later—this time along with Wiener and Oystein Ore, an Oslo native.

†*Atti Congr. Intern. Mat. Rome.* vol. i, pp. 179–89; see Veblen's Rice Institute Lectures on Modern Geometry (Rice Pamphlet XXI (1934)), for a more complete exposition of his ideas.

Whitney among others. Four of these became AMS Presidents (Morse, Stone, Walsh and Morrey), and five achieved tenure at Harvard (Morse, Walsh, Stone, Whitney, and Widder), incidentally contributing much to my own education.† Two of them (Morse and Whitney) later transferred to the Institute of Advanced Study, while a third (Stone) revamped the Mathematics Dept. of the University of Chicago after World War II. Though all eight of those listed had distinguished careers, the most influential during the decades 1921–41 were Morse and Stone.

After getting his Ph.D. in 1917, Harold Marston Morse (1892–1977) served in World War I, first in the Ambulance Corps and then in the Artillery. When he returned, he first studied (at Cornell) the topological structure of the set of geodesics on a general closed surface. Then, around 1924, he began laying the foundations of what is today called Morse theory.‡ Its original stimulus was the minimax principle of G. D. Birkhoff, which Morse extended into a celebrated series of "Morse inequalities" relating the numbers of critical points of specified type that a function could have on a global manifold to the topology of the manifold. During the next decade, Morse developed his methods further into a "calculus of variations in the large," written up as an AMS Colloquium volume (1934). Bochner has described†† how this then new topological approach was quite beyond the grasp of Carátheodory, illustrating the extent to which American mathematics was forging ahead by this time.

While at Harvard, Morse supervised the Ph.D. Theses of G. A. Hedlund, M. H. Heins, Herbert Robbins, S. S. Cairns, Walter Leighton, and Arthur Sard (besides stimulating me as an undergraduate). Shortly after his Colloquium Lectures were published, Morse left Harvard for the Institute for Advanced Study, where he influenced R. H. Fox, Richard Arens, Pesi Masani, and James Jenkins among others.

Stone. The first published research of Marshall Harvey Stone (1903–   ) dealt with series expansions in the tradition of Bôcher. Around 1928, he turned his attention to the theory of linear operators on an abstract Hilbert space, simultaneously with von Neuman. He soon proved several sharp new theorems, of which his spectral resolution of one-parameter unitary groups is most famous. He incorporated the contempory work of von Neumann (*q.v.*) on self-adjoint operators, a generalization of the Hahn-Hellinger unitary equivalence theory, and many new results of his own about differential operators into the first major treatise on operator theory, entitled *Linear*

---

†Since all are very much alive, I hope they will excuse the temerity of my summary remarks.

‡See TAMS 27 (1925) 345–96; 30 (1928) 213–74; 31 (1929) 379–404; and 32 (1930) 599–631; also Math Annalen 103 (1930) 52–69.

††S. Bochner, Amer. Math. Monthly 81 (1974) 827–52. Since Carátheodory's *Variationsrechnung* is a profound and highly respected treatise, Bochner's recollection is highly significant.

Transformations on Hilbert Space (Amer. Math. Soc., 1930). This and Banach's *Théorie des Opérations Linéaires* (Warsaw, 1933) were the bibles of the functional analysts of the 1930's.

By 1933, Stone was breaking new pathways in general topology (point-set theory) and Boolean algebra, a subject which had previously been studied mostly in connection with logic and postulate theory. He was the first to rigorously subsume the theory of Boolean algebras under the general theory of rings, (a "Boolean ring" is just a ring with unity in which every element is idempotent), by establishing a simple 'cryptomorphism' between Boolean algebras and Boolean rings. He then established a much deeper cryptomorphism between Boolean algebras and one-dimensional compact spaces (today called "Stone spaces").

J. L. Walsh. Another distinguished Ph.D. of G. D. Birkhoff was Joseph Leonard Walsh (1895-1973). Though a contemporary of Morse, his career and mathematical style were very different. He taught at Harvard from 1916, when he taught Freshmen as a "section man" under the plan that had just been organized by the W. L. Putnam Committee (see §12), until 1966, when he went to the University of Maryland. A native Marylander, he was a vigorous man of outstanding courage who served as deck officer in two world wars. He also volunteered for duty as a policeman during a month long police strike in the 1920's, patrolling one of Boston's more dangerous districts.

For over 50 years after returning from World War I, Walsh devoted his energies to classical analysis, teaching complex analysis in the spirit of Weierstrass and Osgood. The orthogonal "Walsh functions" with range ± 1, constant on binary intervals, were one of his first inventions (Amer. J. Math. 45 (1923) 5-24). But the bulk of his work was on interpolation and approximation theory; an important early result was that any function harmonic on a compact domain can be uniformly approximated arbitrarily closely by harmonic polynomials. Runge had proved a corresponding extension of the Weierstrass approximation theorem to complex polynomials two decades earlier.

Although approximation theory had been of interest to Cauchy, Hermite, Chebyshev, Weierstrass, Hadamard, and de la Vallée-Poussin, the New Mathematicians of the 1930's (see §20), 1940's and 1950's had little time for such special results.† As a result, approximation theory became less and less fashionable during these decades, and it was not until the late 1960's that the value of Walsh's research became generally appreciated (see §23).

Both Stone and Walsh had many distinguished students. To mention only a few of the most outstanding, Stone supervised the theses of George Mackey, John Calkin, Edwin Hewitt, H. MacNeille, R. V. Kadison, and Bernard

---

†Cf. J. Dieudonné, Math. Revs. 12 (1950) 249-50.

Galler (later President of the Association for Computing Machinery); while Walsh supervised those of Morris Mardin, J. L. Doob, M. H. Heins, L. H. Loomis, R. S. Varga, and I. E. Block (a founder of the Society for Industrial and Applied Mathematics and longtime mainstay of its publications).

16. Norbert Wiener. Norbert Wiener (1894-1964) became one of the most original and stimulating American mathematicians of his generation. His two autobiographical books [22] and [23] give a candid and vivid picture of his career, and of the contemporary mathematical scene as well. Unfortunately, he was too often spiteful, and these books must be corrected for astigmatism. Fortunately, the sympathetic but honest biography of Wiener in [24, pp. 1-32] by Norman Levinson, his best student, gives the necessary correction. (See also [29, p. 92 ff].)

Trained by his father to be a prodigy, Wiener's first mathematical venture was into mathematical logic. After graduating from Tufts College at fourteen he entered the Graduate School of Harvard University, where his father was a distinguished (if mildly eccentric) professor of Slavic languages. Abbott Lawrence Lowell, by coincidence, became Harvard's President that same fall (1909).†

After a year trying his hand unsuccessfully at zoology and botany, then a year at Cornell (to which he went from Boston via interurban trolley!) Wiener returned to Harvard as a 16 year old graduate student in philosophy. Here he became interested in postulate theory by his father's friend E. V. Huntington, and his interest led to a Ph.D. in mathematical logic in 1913.

He then went to Cambridge University, intending to study with Bertrand Russell (1872-1970), the co-author with A. N. Whitehead of the classic treatise on symbolic logic, *Principia Mathematica*. There he also met G. H. Hardy (1877-1947), the great British analyst and analytic number theorist, "the mathematician who was to have the greatest influence on me in later years."

Wiener went through many curious wartime experiences, which he describes at length in [22]. Most relevant is his stay at the Aberdeen Proving Ground [22, pp. 254-63 and 294-5], where he met Veblen, Bliss, Gronwall, Alexander, Ritt, and Bennett, as well as his future brother-in-law, Philip Franklin of M.I.T.**

M.I.T. Wiener joined the M.I.T. faculty in 1919. The next summer he

---

†Wiener's distorted description of Lowell [22, pp. 125-6] is frightening! His malicious caricatures of Felix Klein [22, pp. 209-10] and [23, pp. 96-7], Osgood [22, pp. 231-2]. G. D. Birkhoff [23, pp. 27-28] and Courant [23, p. 96 and pp. 114-6] are in the same vein, as is his verdict that Landau "had neither taste nor judgment nor philosophical reflection."

** Had it not been for the 1919 influenza epidemic, Wiener's brother-in-law would have been Harvard mathematician Gabriel Marcus Green instead!

went to Europe (for the fourth time; he would go again in 1922, 1924, and 1925), to attend the International Mathematical Congress in Strasbourg. His impressions of this [**23,** pp. 66-68] make lively reading; thus he describes Camille Jordan reminiscing during group walks about "the great days when Cauchy was lording it over French mathematics and forcing all the younger men to pay tribute." (In the 1920's and 30's this role was played by Picard!)

Before this Congress, Wiener spent some weeks following Fréchet around. Though he was not attracted by Fréchet's topological foundations of functional analysis, he was inspired to define the notion of a Banach space at the same time as Banach [**23,** p. 60]. Curiously, in their search for extreme generality neither Fréchet nor E. H. Moore had ever formulated the simple concept of an abstract Hilbert space. F. Riesz had likewise collected together the defining properties of a Banach space, † but had not given a name to linear spaces satisfying them. Fréchet also introduced Wiener to the probabilist Paul Lévy, whose ideas about Brownian motion Wiener surely found stimulating.‡

At M.I.T., Wiener finally found a congenial environment. Within five years, he had written two important papers on the Dirichlet problem under the influence of H. B. Phillips and O. D. Kellogg,†† and (most important) had constructed a rigorous measure-theoretic model for Brownian motion. Apparently Wiener's attention was called to the problem of constructing measures in function spaces by I. A. Barnett, a Ph.D. of Bliss who was familiar with the E. H. Moore tradition.‡‡

Wiener then extended his ideas to prove rigorously that the autocorrelation function, introduced by G. I. Taylor* to describe turbulence, was the Fourier transform of the energy density. The connection is explained in Wiener's famous 1930 paper on 'generalized harmonic analysis.' Three years later, he wrote a related paper on 'Tauberian Theorems' (Annals of Math. 33 (1932) 1-100), in which he proved results more powerful and general than those of his teachers Hardy and Littlewood. For this work, he shared the Bôcher prize with Morse in 1933.

Fortunately, Wiener's many-sided original contributions have been carefully reviewed in [**24**].** I shall only mention further the Wiener-Hopf theory, and his determination (with Paley) of which transfer functions (gain and

---

†Acta Math. 41 (1918) 71-98.

‡See J. P. Kahane in [**9**, pp. 595-600], where reference is made to Paul Lévy's autobiography.

††J. Math. Phys. MIT (1923) 105-24, and 3 (1924) 127-47. The first paper, like that of Richardson, exploited the 5-point difference approximation to the Laplace equation.

‡‡[**22**, p. 274] and [**23**, p. 174]. The most relevant previous paper was by Fréchet, Bull. Soc. Math. France 43 (191) 249-67. Fréchet's student Gateaux was killed in the war.

*Proc. Lond. Math. Soc. 20 (1921) 196-212; Wiener's paper was published in Acta Math. 55 (1930). Wiener refers to Taylor, a leading scientist of his time, as an "amateur with a professional competence" [**23**, p. 37].

**See also Wiener's *Selected Papers*, M.I.T. Press, 1964.

phase lag) are "realizable" or *predictive* (i.e., make the output depend only on *past* input).

Wiener was also notable as one of the few Americans of his time who was outstanding in both pure mathematics and his applications. How much of this can be attributed to his varied and cosmopolitan early background, and how much to his continuing contacts with non-mathematicians such as G. I. Taylor, it is hard to say. But it is clear that he interacted most fruitfully with engineers on the M.I.T. faculty. Thus Harold Hazen and others built in 1930 an instrument for measuring the autocorrelation function

$$R(\tau) = \overline{x(t)x(t + \tau)} \, / \, \overline{x^2(t)},$$

used by Wiener to define the *spectrum* of a function.

Bush. Another electrical engineer with whom Wiener interacted fruitfully was Vannevar Bush. An expert in electrical circuit theory, Bush liked to characterize himself as a Yankee inventor. His most notable invention was the differential analyzer, an electromechanical analog machine that solved systems of ordinary differential equations with 3 digit accuracy. Much as later electromechanical and electronic digital computers can be regarded as American technological implementations of ideas of Babbage, so the Bush differential analyzer was a more practical realization of ideas of Kelvin and his brother James. For a fuller account I refer you to Bush's 1935 Gibbs Lecture, to its 1966 sequel by Derrick Lehmer,‡ and to Wiener's remarks in [**23**, pp. 136–9].

In spite of the great importance of Wiener's personal contributions to applied mathematics, none of his students seems to have become an important industrial mathematician. In constrast, at least three of G. D. Birkhoff's Ph.D.'s did: Joseph Slepian, the inventor of the magnetron, was prominent at Westinghouse; M. H. Slotnick became research director for the Humble Oil Co.; and H. Poritsky was an active research worker at General Electric.

17. Applied mathematicians. During the 1920's and 1930's, few American mathematicians paid much attention to contemporary engineering developments. Although they used the new concepts of relativity and quantum mechanics as stimuli for mathematical speculations, the solution of special problems bored them. This led to a paradox: while American industry was growing by leaps and bounds, our academic mathematicians were taking less and less interest in it. Thus Thorton Fry [**11**, p. 10] estimated that in

---

‡Kelvin's original papers appeared in Proc. Roy. Soc. A24 (1876) 262–75; his ideas were built into a tide predictor by our Coast and Geodetic Survey. The Gibbs Lectures cited were published in the Bull. AMS 42 (1936) 649–70, and 72 (1966) 739–50.

1940 only about 150 mathematicians were employed in American industry, "most of whom were trained as physicists or (electrical or mechanical) engineers, but had gravitated into their present work because of a strong interest in mathematics".

Nevertheless, a few American-trained men did make notable contributions to applied mathematics during the decades 1911–41. I have already mentioned E. W. Brown; I shall now mention some others.

E. B. Wilson. A most successful and versatile applied mathematician was Edwin Bidwell Wilson (1888–1964). After graduating from Harvard in 1899, Wilson went to Yale on Osgood's advice. There he discovered J. Willard Gibbs, whose eminence was recognized by B. O. Peirce, but not (apparently) by Osgood.† He helped Gibbs write his famous *Vector Analysis*, which expounded Gibbs' ideas about vectors and matrices ("dyadics"), and made Hamilton's quaternions obsolete almost overnight.

He then went to M.I.T., where he was assigned the job of organizing the first course in aeronautics ever taught at M.I.T., if not in our country. This was only a few years after the Wright brothers made their first successful flight, and only a decade or so after Simon Newcomb was arguing that airplanes were a scientific impossibility.*

Wiener mentions the encouragement he got from E. B. Wilson, then a professor of physics, when he first went to M.I.T. in 1920 [**20**, pp. 71–72]. Not long after, Wilson moved to the Harvard School of Public Health as an applied statistician. Among his many other distinctions, he served as Managing Editor of the *Proceedings* of the National Academy of Sciences from its inception in 1915 until 1964, when he was succeeded by Saunders Mac Lane.

Statistics. It is hard to realize today, when many Divisions of the Mathematical Sciences include Departments of Applied Mathematics, Statistics, and Computer Science, how little academic recognition was given to any of these fields in the 1920's.

Thus H. L. Rietz of the University of Iowa was one of the very few mathematical statisticians in the 1920's;‡ there were only actuarial statistics, agricultural statistics, economic statistics, educational statistics, and public health statistics, sharing a very loose bond of common theory. Even in the 1930's, the development of quality control by Shewhart took place at the

---

†See E. B. Wilson's "Reminiscences of Gibbs," Bull. AMS 37 (1931) 401–16. On pp. 414–5 of this article, Wilson criticizes the exclusive concern of the AMS establishment with pure mathematics.

*Even in the early 1920's, I remember G. D. Birkhoff gazing in admiration when he heard a plane flying overhead, in much the same way that people gazed at the first Sputnik in 1959.

‡In the Preface of his *Mathematical Statistics.* he says "considerable portions of this monograph can be read by those who have relatively little knowledge of college mathematics."

Bell Telephone Labs., not at a university, and the work of such mathematical statisticians as Harold Hotelling and S. S. Wilks went almost unnoticed by most mathematicians.

Mason and Weaver. Two other important figures in American applied mathematics were Max Mason (1877–1961) and Warren Weaver (1894–), two friends who co-authored a distinguished book *The Electromagnetic Field*. In some ways, the careers of these men of action recall those of the first AMS presidents.*

Mason wrote his Ph.D. Thesis on boundary value problems with Hilbert. After giving the AMS Colloquium Lectures in 1906 (with E. H. Moore), Mason turned to physics but did not publish. However, he showed his mettle again in World War I by developing an extremely effective submarine detector called the MV-tube. A few years later, he became President of the University of Chicago; then head of the Natural Sciences Section of the Rockefeller Foundation, where Weaver succeeded him in 1932.

Shortly after being appointed, with the full backing of Mason, Weaver persuaded the Foundation to actively support "the application to basic biological problems of the techniques, experimental procedures and methods of mathematics so effectively developed in the mathematical and physical sciences."** Indeed, the Natural Sciences section of the 1938 Annual Report (drafted by Weaver) *began* with a 16 page section headed Molecular Biology. This was at a time when most mathematicians thought of genetics in terms of Mendel's laws, and 15 years before DNA and RNA revolutionized genetics.

Continuum mechanics. During the half-century 1891–1941, our country was especially backward in continuum mechanics; almost all our leading experts were emigrés from Europe. Thus even before 1900, the British-educated astronomer E. W. Brown was chosen to review Lamb's classic *Hydrodynamics*.† And when the National Research Council sponsored a review of fluid mechanics in 1932, the bulk of the text was supplied by two other British-trained experts, F. D. Murnaghan (1893–1976) and H. Bateman.‡

The situation was similar in solid mechanics: S. Timoshenko, a Russian emigré who went to Stanford, was our most influential figure; another was Ivan Sokolnikoff (1901–76), also a Russian emigré. Murnaghan tried to

---

* See [25], vol. 37 (1964) 205–36 for Weaver's biographical sketch of Mason; Weaver has published his own reminiscences in [20].
** Quoted from a letter from Warren Weaver.
† See Bull. AMS 4 (1897) 73–89. Lamb was actually a Professor of Mathematics; his book went through six editions.
‡ H. L. Dryden, F. D. Murnaghan, H. Bateman, *Hydrodynamics.* Bull. Nat. Res. Council, 1932; Dover reprint, 1956. Dryden (1898–1965) was a Physics Ph.D., but contributed only 28 pages out of 625!

construct workable mathematical models of nonlinear elastic solids which were consistent with Bridgman's high-pressure data,* but none of these men attracted mathematical disciples of comparable stature.

The contrast with Europe is striking;** Felix Klein's influence was again strong, and his protegé L. Prandtl (1875-1953) developed a whole school in fluid mechanics. T. Levi-Civita (1873-1941), famous also for his work in differential geometry and on the three-body problem, was another notable European contributor to applied mathematics. Together with Th. von Kármán (1881-1963), R. von Mises (1883-1953), G. I. Taylor, S. V. Southwell, D. M. Burgers, S. Goldstein, and many other European scientists and engineers, these men organized a series of international congresses in "theoretical and applied mechanics" from 1922 on at which Germans and Austrians were welcome, but whose *Proceedings* were largely ignored by American mathematicians.

THE UNITED STATES ASSUMES LEADERSHIP

18. John von Neumann. The mathematical ascendancy achieved by our country in the 1930's owes much to John von Neumann (1903-1957), one of the most scintillating minds of this century. Born into a well-to-do Budapest banking family, he was always brilliant, getting advanced degrees in both chemical engineering and mathematics in 1926.† By that time, he had already proposed new axiomatic foundations for set theory.

During the years 1926-29, while a Privatdozent at the University of Berlin, he published 25 more papers, several of them fundamental. In particular, he developed his axiomatic foundations of set theory further, made an excursion into Hilbert's proof theory, and showed that the existence of a non-trivial additive measure on Euclidean n-space, invariant under the Euclidean group, depends on the *solvability* of that group. Even more important, he showed how to construct a spectral resolution for unbounded Hermitian operators on an abstractly defined Hilbert space, using new theoretical concepts going beyond those of F. Riesz. And finally, in a series of papers beginning with a joint 1927 paper with Hilbert and L. Nordheim, he showed that these concepts provided a natural and mathematically rigorous setting for the then new theory of quantum mechanics. He was still only twenty-six!

---

*Richter of "Richter scale" fame refers to Murnaghan's work in his Gibbs Lecture (Bull. AMS 49 (1943) 478-93).

**Timoshenko has given in [18, Chs. XIII-XIV] a charming and informative account of the work of the leaders in solid mechanics; many of them also worked in fluid mechanics.

†For these and other facts and legends about von Neumann, see Paul Halmos, Amer. Math Monthly 80 (1973) 382-94; also [2] and [29].

In 1930, he married and went to Princeton, which was to remain his home for the rest of his life. Two years later, his famous *Grundlagen der Quantenmechanik* presented his reformulation of quantum mechanics in book form* and made so-called 'operator theory' fashionable overnight. This book has served as an inspiration and model for mathematicians interested in the foundations of quantum mechanics ever since.**

In the meantime, he had begun interacting with leading American mathematicians. Osgood's nephew B. O. Koopman had observed that any one-parameter group of measure-preserving transformations (e.g., the flows in phase-space associated by Liouville's Theorem with dynamical systems) defined a one-parameter unitary group to which Stone's theorem applied. This stimulated von Neumann to formulate and prove the Mean Ergodic Teorem. It was his discussion of this with G. D. Birkhoff that stimulated the latter to prove his sharper Pointwise Ergodic Theorem (see [6, vol. ii, pp. 462–5]). That same year, von Neumann also sharpened Stone's theorem on one-parameter groups of linear isometries of Hilbert space; returning the compliment, Stone's *Linear Transformations in Hilbert space* gave a more systematic and didactic treatment of the whole subject. At 29, von Neumann had established himself as one of the world's leading younger mathematicians—but he still published in German, indicating his intention to return to Europe, possibly to Göttingen.

The next year, Hitler seized control of Germany, the Institute for Advanced Study was founded in Princeton, and von Neumann became one of its first professors. He also became co-editor of the Annals of Mathematics; its founder Ormond Stone had died the year before, an associate editor until the end. The die was cast;† von Neumann began publishing in English, and instantly became a leading *American* mathematician.

It seems almost superfluous to touch on his outstanding research contributions of the next eight years. Already in 1933, he solved Hilbert's Fifth Problem for compact groups, by proving that any compact locally Euclidean topological group was topologically isomorphic to an analytic group (a Lie group). He then deepened his analysis of the structure of 'operators' (linear Hermitian operators on Hilbert space), founding in this connection a new theory of rings of operators (with F. J. Murray), and making major contributions to lattice theory, then enjoying a renaissance. But above all, he was

---

*English translation, Princeton Univ. Press, 1955. Gossip has it that Springer made von Neumann write his ideas up in book form, to pay for enormous charges for changes in page proofs of previous papers.

**See George Mackey, "The Mathematical Foundations of Quantum Mechanics," Benjamin, 1963. The author wishes to thank Professor Mackey for many helpful criticisms of an earlier draft of this paper.

†Von Neumann's decision to emigrate from Europe is also discussed in [29, p. 68]. For a review of von Neumann's influence, see Bull. AMS, vol. 64.

interacting with American mathematicians of all ages, helping to set the mathematical style of the 1930's which I shall discuss next.

By 1940, he was actively preparing himself for war research, and had already published his first paper with R. H. Kent (see §22, end). This led to even more brilliant post-war activity, concerned with computing machines and numerical mathematics, but this is outside the scope of my talk.

19. The early 1930's. In the early 1930's, the world was in the depths of the Great Depression. The job situation was so bad in our country that the MAA appointed a Commission in the summer of 1933, with E. J. Moulton as Chairman, on the Training and Utilization of Advanced Students in Mathematics. It was found that "of 3488 mathematics teachers in the colleges of the U.S. and Canada, only 937 held doctorates."‡ Of these probably less than 150 were active in research. Nevertheless, many Ph.D.'s were unable to get jobs in colleges, and (like Maschke in Germany 60 years earlier) had to accept high-school positions; even getting these was not easy.

Institute for Advanced Study. A new bright spot in these gloomy years was provided by the Institute for Advanced Study in Princeton, endowed by the generosity of Felix Bamberger, a phenomenally successful Newark department store owner. Though it did not alleviate the unemployment problem, it was designed to provide ideal conditions in which outstanding scholars could work and radiate their wisdom. Its design followed specifications outlined by Abraham Flexner (1866-1959) in his influential book *Universities: American, British, German.*

It began with a School of Mathematics, whose early members included Veblen, von Neumann, Morse, and Einstein and Weyl from Europe. During the 1930's, it both enabled many young Americans to do research in a stimulating environment, free from teaching duties, and (later, see §21) provided a haven for scholars fleeing Nazi-dominated portions of Europe.

Society of Fellows. A second bright spot was Harvard's Society of Fellows (see §12), founded on very different principles. It was designed to *free* outstanding young men to do research, explicitly emancipating its members from Ph.D. requirements. Membership in this played an important role in the careers of R. C. Buck, Gleason, Loomis, Mazur, Minsky, Mumford, Oxtoby, Quine, Ulam, and G. Birkhoff—not to mention three Nobel Prize winners (Bardeen, Samuelson, Woodward), one a double winner. Lowell's idea has since been copied at other universities.

Veblen and G. D. Birkhoff still headed our Mathematical Establishment,

---

‡Amer. Math. Monthly 42 (1934) 143-4. Over half of these were from Chicago, Cornell, Harvard, Illinois, Johns Hopkins, or Yale.

in friendly cooperation with Bliss, Dickson, R. G. D. Richardson, and many others. However, G. D. Birkhoff was to serve as Dean of the Faculty at Harvard for four years, 1934–38, while Veblen was to become the guiding force of the School Mathematics at the Institute for Advanced Study.* In spite of these administrative distractions, which greatly reduced their time for creative mathematical thought, they both attended AMS meetings regularly and continued to think of mathematical research as their highest activity.

In the eyes of the mathematical community at large, Analysis still appeared to reign supreme. Thus the first two Fields Medals were awarded (in 1936) to two analysts: Ahlfors for his work on the geometry of Riemann surfaces, and Douglas for his solution of the Plateau problem, in competition with Courant and Tibor Rado. Douglas was at Columbia, and Ahlfors at Harvard at the time; hence these awards enhanced America's mathematical prestige.

20. New trends of the 1930's. However, Analysis in the traditional sense was no longer the central concern of the most active younger American mathematicians of the 1930's. Although we acknowledged the preeminence of such older analysts as G. D. Birkhoff and Hermann Weyl, young men like A. A. Albert, Nelson Dunford, Nathan Jacobson, Saunders Mac Lane, Deane Montgomery, Barkley Rosser, Norman Steenrod, John Tukey, and myself tended to regard logic, abstract algebra, topology, and functional analysis as more promising areas for important future discoveries. † Perhaps this was partly because we felt we could not compete successfully with the masters on their own territory.

As a result, we concentrated our attention on the current research of men like von Neumann, Stone, Pontrjagin, and Whitney. When Artin and Chevalley joined Wedderburn at Princeton, this trend accelerated. We found excitement in the dramatic discoveries of Gödel in logic, in the sweeping generalizations made by Emmy Noether and her school in algebra, and in the work of Élie Cartan and O. Schreier on topological groups. These opened new vistas extending far beyond the structure theorems for associative and Lie algebras that had been obtained by Cartan and Wedderburn 30 years earlier. Inspired by the recent successes of von Neumann, Stone, Schauder, and Leray, we also hoped that the new theories of linear operators on Hilbert and Banach spaces would lead to enormous generalizations and simplifications of the classical theories of differential equations. Indeed, our hopes

---

*For the record, it should be stated that G. D. Birkhoff was invited to join the IAS at a salary of $20,000 (equivalent after taxes to over $80,000 today), but declined to leave Harvard. Institute salaries were then set at $15,000. (Harvard's range for full professors was $9,000–$12,000, truly luxurious during the Great Depression.)

†E. J. McShane, Norman Levinson, C. B. Morrey, and other young analysts naturally did not share this view.

were to be in some measure fulfilled, although it was to be the Bourbaki school in France that would give them their most definitive expression.

Above all, we hoped to be free to pursue our researches into fundamental questions of pure mathematics suggested by these new ideas and techniques, and to become famous by discovering still other basic new ideas and techniques, undisturbed by the political storms which had already driven so many European colleagues to our shores in search of asylum.

I think that the preceding remarks correctly reflect the mood of our country's leading younger mathematicians during the years preceding Munich. As the semicentennial anniversary of the founding of the AMS approached, most American mathematicians were justifiably proud of the progress and momentum achieved during the preceding fifty years. This mood of patriotic pride is very evident if one reads the two volumes published to commemorate this anniversary ([1], [2]). The younger speakers chosen include McShane (best student of Bliss), T. Y. Thomas (best student of Veblen's differential geometry phase), and R. L. Wilder (best student of R. L. Moore). The other technical lectures were by Evans, Ritt, Wiener, and Synge—Synge presumably being chosen to represent both the British tradition in applied mathematics[†] and Canadian participation in the AMS.

G. D. Birkhoff's lecture [5] on "Fifty Years of American Mathematics" gives perhaps the most authentic technical account of progress in American mathematics during this period, especially when read in conjunction with Morse's account of G. D. Birkhoff and his work in [6, vol. i, pp. xxii-lvii][‡]. To round out the picture, one should consult the scientific biographies of 17 leading American mathematicians "in their thirties" published by *Scripta Mathematica* in vol. 4, pp. 87-93, 188-95, 283-9, and 330-4.[*]

In this connection, as a final example of a "leader in American mathematics" during the period under discussion, I want to cite Hassler Whitney, who exemplified our desire for new ideas and new problems.[**]

Hassler Whitney. Simon Newcomb's grandson Hassler Whitney (1907-) was the most independent and self-motivated of G. D. Birkhoff's post-1930 students. I well remember G. D. Birkhoff's annoyance when Whitney's first minor thesis, a piece of expository writing required of all Harvard Ph.D. candidates, was turned down by the Harvard Mathematics Department (with Osgood as chief critic) for not being a careful enough exposition!

Whitney's doctoral thesis was on the four color problem, for which he

---

[†]I hasten to add that, like Hamilton (and Murnaghan), Synge is Irish.

[‡]Reprinted from Bull. AMS 52 (1946) 356-91.

[*]Of the 13 "native sons," 10 were Ph.D's of G. D. Birkhoff, Bliss, Dickson, Veblen, or R. L. Moore—hence in the E. H. Moore family group.

[**]One could also mention A. A. Albert (1905-72); see N. Jacobson, Bull. AMS 80 (1974) 1075-1100, and D. Zelinsky, Amer. Math. Monthly 80 (1973) 661-5.

found an equivalent graph-theoretic formulation. In turn, this led him to formulate an abstract theory of linear dependence which inspired, much later, a great deal of combinatorial research.

But his most important contribution was to the formulation of the new algebraic concepts underlying the algebraic topology that was beginning to take shape during those years. He was a major contributor to the recognition and formulation of such notions as 'fiber bundle', 'sphere bundle', 'cocyle', 'coboundary', 'differentiable manifold', and 'tensor product'. In a few years, he would become distracted by war work, but it was to no small extent Whitney's prewar ideas that were to make topology progress in new directions after 1940.

Lefschetz and Tamarkin. The new approaches to topology (to which Steenrod and Tukey were contributing actively) are expounded in Lefschetz' *Algebraic Topology* (1942),† and it is typical of the revolutionary spirit of the times that Lefschetz' personal decision to rename "combinatorial topology" "algebraic topology" was accepted virtually without dissent. (Today, combinatorics is staging a vigorous comeback!)

It is also noteworthy that J. D. Tamarkin helped read the proofs of this influential book. Though an analyst, Tamarkin's personal support and stimulus were felt in every phase of the American mathematical effort. To take one example, he personally translated Ado's fundamental paper on Lie algebras for my benefit!

21. Europe in torment: 1914–41. There is no doubt but that our nation became the world's leading mathematical country during the 1930's, if not before. However, this status owed much to our political and economic stability, as contrasted with European instability.

I have already mentioned (in §8) the tragic impact of World War I on European mathematics. The loss of life was probably felt most deeply in France and Great Britain, which had relatively low birth rates; but postwar upheavals continued to distract Russia and Germany; while the old Austro-Hungarian empire was Balkanized into small fragments. Only Holland, Switzerland, and Scandinavia emerged unscathed, although Polish independence stimulated a remarkable resurgence of mathematics there: overnight, Poland became the world center of point-set topology (set theory).

Our country (like Western Europe) was enriched by the resulting flow of Russian refugees: J. D. Tamarkin, J. Shohat, Sokolnikoff, Timoshenko, Uspensky, Zariski, and others added substantially to our mathematical resources. In return, we tried to help in reconstructing Western Europe; the establishment of the Institut Henri Poincaré in Paris and of a new Mathe-

---

†See the Preface, which mentions Whitney four times.

matical Institute in Göttingen, both financed by the Rockefeller Foundation (see §14), being notable instances.

It was not until the 1928 Congress at Bologna that nationals from the defeated countries of Germany, Austria, Hungary, and Bulgaria were allowed to participate in an international mathematical congress, after a lapse of 16 years! Mussolini had stabilized Italy politically six years before,† and nobody had yet challenged the League of Nations.

Indeed, at the 1928 and 1932 International Mathematical Congresses (the latter at Zurich), one had the impression of a still dominant Europe; see the attendance statistics below:

	1928	1932	1936
England	47	37	48
France	56	69	28
Germany	76	118	35
Italy	336	64	5
Russia	37	10	11
U.S.	52	66	86
Poland	31	20	25
Scandinavia	21	20	101
Low Countries	19	23	24
Austria & Hungary	31	22	15
Czechoslovakia, Yugoslavia, Rumania	30	23	21
	836	667	487

However, there were deep-seated resentments associated with economic dissatisfaction. In France, bus-loads of American tourists were stoned in the mid-1920's; in England, laborites were angry at 'old school tie' favoritism and economic privilege. Finally, Hitler capitalized on bitterness over the Treaty of Versailles and the post-war inflation to seize power in Germany by a classic coup d'état in March 1933. Making communists and wealthy Jews his first targets, he started a campaign of pan-German expansion.

At the time, owing to the emigration of a small number of outstanding mathematicians such as Hille, Ore, E. Hopf, T. Rado, D. J. Struik, A. Wintner, and especially von Neumann, our country had already begun to achieve a mathematical strength commensurate with its population and resources. But after 1933 there came to our shores a veritable flood of new refugees from Germany and neighboring countries menaced by Hitler. As a result, whereas Göttingen was still acknowledged to be the world's greatest

---

†Incidentally, Italian hydroelectric power had doubled during those years, from 3.65×10^6 Mwh to 7.8×10^6 Mwh annually. See the first paper in the Congress *Proceedings*!

mathematical center by most mathematicians (especially those at Göttingen!) in 1928, it was a shambles by 1936. The *Zentralblatt für Mathematik,* Springer's progressive rival to the hoary *Jahrbuch der Fortschritte der Mathematik,* was being run from Copenhagen by Neugebauer, and our country had become the symbol of hope for the future. Although the 1936 International Mathematical Congress in Oslo took no official notice of these unpleasant facts, it was not surprising that an invitation to hold the 1940 International Mathematical Congress in Cambridge, U.S.A., with G. D. Birkhoff as President, was accepted with thanks. Or that G. H. Hardy should have stated in his after-banquet speech at the 1936 AMS Summer meeting in Cambridge, that the United States had become the world's leading mathematical country.

Storm clouds were visible all around, but almost everyone still hoped in 1938, at the time of the Munich agreement, that World War II could be avoided.

22. Twilight of an era. The semicentennial anniversary of the AMS was celebrated weeks after Munich. During the years 1938–41, events came thick and fast. In particular, "American" mathematics lost its indigenous character.

Even before 1938, the migration of eminent European mathematicians to our shores had become a deluge. Some of the names that spring to mind are: Artin, R. Brauer, E. Noether, Schilling; Courant, Friedrichs, Lewy; Bochner, Pólya, Szegö, Rademacher, Schoenberg, Weinstein; Eilenberg, Hurewicz, Kac, Ulam; Wigner, Menger; Chevalley, André Weil.**

Another important acquisition was O. Neugebauer, the distinguished historian of mathematics, who transplanted the *Zentralblatt für Mathematik,* by a painless mutation, into an English language *Mathematical Reviews* published by the AMS. The AMS headquarters followed Neugebauer to Providence a decade later.

Hermann Weyl. Of all the mathematicians who emigrated to the United States in the 1930's, Hermann Weyl (1885–1955) was the most eminent. The successor of Hilbert and Felix Klein at Göttingen, his move to the Institute for Advanced Study in 1933 had a symbolic as well as a real significance.

Weyl was a contemporary of G. D. Birkhoff, and their early research was similar. Thus both were deeply interested in expansions in eigenfunctions, and more broadly in the differential equations of mathematical physics, using integral equations as a tool.† However, whereas G. D.

**For a more complete list, see A. Dresden, Amer. Math. Monthly, 49 (1942) 415–29.

†See H. Weyl, Bull AMS 56 (1950) 115–39; for Weyl's own hindsight on the half-century 1900–1950, see Amer. Math. Monthly 58 (1951) 523–33. For a personal appraisal of Weyl and some of his work, see C. Chevalley and A. Weil, Ens. Math. 3 (1957), reproduced in [18, vol. iv, pp. 655–85].

Birkhoff concentrated on *ordinary* differential equations and dynamical systems having a finite number of degrees of freedom, Weyl branched out early into *partial* differential equations and complex analysis.

One of Weyl's spectacular early results was his proof that all Dirichlet-type problems had the same asymptotic distribution of eigenvalues λ_j, the number of $\lambda_j \leqslant \Lambda$ being asymptotically proportional to $\Lambda \cdot \text{vol }(R)$. Another early influential contribution to the philosophy of Riemann surfaces was his monograph *Das Idee der Riemannsche Fläche*, enthusiastically dedicated to Felix Klein. This contained an early rigorous formulation of the 'great and divine' ("gross und gottlich") concept of a Riemann surface, and expounded Koebe's uniformization theorem with the global rigor of L. E. J. Brouwer.

During the tragic years of World War I, Weyl wrote his influential *Raum, Zeit, Materie,* a book whose English translation† (after four German editions) helped greatly to clarify the mathematical meaning of Einstein's special and general theories of relativity. In the preface, Weyl mentions its "intermingling of philosophical, mathematical, and physical thought, a study which is dear to my heart." Weyl's philosophical bent was even more forcefully expressed in his attacks on Hilbert's "formalist" approach to mathematics in the early 1920's; Weyl advocated the "intuitionist" approach of Brouwer.

But most important were his contributions to group representation theory in the mid-1920's. Where earlier authors had only determined the representations of finite groups, Weyl determined those of the orthogonal and unitary Lie groups of greatest importance for quantum mechanics, and those of the symplectic group as well. In 1927, he and Peter proved that any compact Lie group can be directly decomposed into finite-dimensional irreducible representations, paving the way for von Neumann's 1935 proof that any compact locally Euclidean continuous group was analytic. When he came to our country, he brought with him a substantial piece of European scientific tradition, and from him flowed wisdom and creative contributions to many aspects of mathematics until his retirement.

During these fast-moving years 1938–41, two parallel currents flowed swiftly side by side. On the one hand, a score or more of able young American mathematicians such as myself strove to make the most of their birthright of freedom and opportunity. For example, it was in these years that I wrote the first edition of "Lattice Theory," and that Mac Lane and I wrote the first edition of our "Survey of Modern Algebra."

There was at the same time also a strong current of European mathematical tradition, trying to adapt itself to a New World in which professors were supposed to be teachers first and publishers of sophisticated research second. The Institute for Advanced Study was a staging ground for many of our distinguished new immigrants; G. D. Birkhoff has discussed some

---

†*Space-Time-Matter* (translated by H. L. Brose) Dutton, 1920.

problems involved in absorbing them in [5, pp. 278-9]. Jobs in stimulating research environments were still very scarce; he feared that if these went to European émigrés, however distinguished, young American mathematicians would become mere "hewers of wood and drawers of water"— Veblen's priorities were the reverse, but solutions were worked out gradually and amicably.

Courant. Another famous refugee from Göttingen was Richard Courant (1888-1972). Like Weyl, he had begun his research before World War I on Hilbert's 'direct' variational methods.* His *Methoden der Mathematischen Physik*, nominally co-authored by Hilbert, was a sensational success among physicists. He came to N.Y.U. from Germany in the middle 1930's, accompanied by two outstanding young compatriots: K. Friedrichs and H. Lewy. Within a decade, Courant had expanded this nucleus into a very large and prosperous research group, specializing in partial differential equations and their application.

Applied mathematics. If pure mathematics was enriched by our newcomers, in applied mathematics they filled a vacuum. Thus the move of von Kármán to the Guggenheim Aeronautical Laboratory at Caltech around 1930 initiated the training of a whole generation of aerodynamic experts, many of whom played leading roles in our World War II effort and our post-war aviation and aerospace industries. Some of von Kármán's skill in applying mathematics can be glimpsed by reading his Gibbs Lecture.†

More profound mathematically were von Mises and his associates: S. Bergman, H. Geiringer, and W. Prager. Together with von Kármán, they imported the sophistication in continuum mechanics that had been lacking in our country up to that time. After the war, von Mises and von Kármán founded the series *Advances in Applied Mechanics*, while Menger's collaborater F. Alt founded *Advances in Computers*.

Another distinguished acquisition was M. Jakob, the world's leading scholar in the area of heat transfer.

After the fall of France in 1940, the urgency of the situation became apparent to all. This made our few native applied mathematicians heroes overnight. T. C. Fry was invited to write an article [11] in the *Monthly*, in which he described the qualities and training that a mathematician in industry must have to be effective, stating that "There is nowhere in America a school where this training can be acquired." R. G. D. Richard-

---

*Math. Annalen 71 (1911) 145-88 and 72 (1912) 517-50, where it is stated that Hilbert's famous vindication of the Dirichlet principle (direct variational methods) was only sketched. Concerning Courant, see [30].

†"The engineer grapples with nonlinear problems," Bull. AMS 46 (1940) 615-83.

son put Fry's ideas into practice a year later, by establishing at Brown a new center where applied mathematics would be taught at the Ph.D. level (cf. Scripta Math. 8 (1941) 57–9).

R. H. Kent. The dawn and the twilight of the era 1891–1941 are linked together by the career of R. H. Kent, Chief Scientist of the Ballistic Research Laboratories at the Aberdeen Proving Ground from around 1920 to 1955. He was among those who learned mathematics and physics from B. O. Peirce, who had been a key figure at Harvard from 1884 on (see §4). Unfortunately for him, but fortunately for our country (see below), Kent failed to get a Ph.D., largely because Harvard required all doctoral candidates in physics to write an experimental thesis before World War I.

Therefore, after inspecting ammunition dumps with J. W. Alexander at the end of that war, he had to take a civil service job at Aberdeen. There he played an inconspicuous but crucial role in maintaining the high technical level established by Veblen and F. R. Moulton, in whose office he served 1917–19 (see [**12**, Part 1, Ch. 9, and Part 2, Ch. 2]). His wisdom and skill in the scientific solution of ordnance problems were material factors in our World War II victory. He also tutored von Neumann,† who had been made a consultant at Aberdeen in 1937, and many other scientists. For his contributions to the war effort, he was elected to the National Academy.

23. Epilogue. The Japanese attack on Pearl Harbor in December, 1941, marked the end of an era. Overnight, the leaders in every phase of American life recognized that there was a national emergency, and that all other objectives had to be subordinated to the defense of our country and institutions.

In a sense, then, my story ends here. But the careers of the leaders I have been describing did not terminate; all that happened was that the advancement of American (pure) mathematics ceased to provide a central focus, and from 1942 on their careers diverged. Let me therefore conclude by briefly recalling how a few of the men whose achievements I have been trying to honor rounded out their careers after 1941.

G. D. Birkhoff, besides continuing his normal duties, toured South America and Mexico as a good-will ambassador representing our scientific culture. It was at the dedication of a new astrophysical observatory in Tonantzintla in Mexico that he announced his relativistic theory of gravitation in flat space-time. He also studied with me the entry into water

---

†They were the authors of two joint papers in 1940–41 (##77–78 of [**15**, p. 532]; see also ##80, 82, 83, 86, and 148). For a sample of Kent's mastery, see Amer. Math. Monthly 48 (1941) 8–14; for Aberdeen's role in World War II, see [**12**, Part 2, Ch. 6].

of air-launched torpedoes; this study led ultimately to my book *Jets, Wakes and Cavities* (1937), whose co-author E. Zarantonello had met G. D. Birkhoff during the latter's wartime visit to Buenos Aires.

Whereas G. D. Birkhoff died in 1944, Veblen lived on for many years. In particular, Veblen took G. D. Birkhoff's place as President of the 1950 International Mathematical Congress in Cambridge, on whose Organizing Committee I had replaced D. V. Widder as Chairman. During the war, Veblen's counsel was also widely sought and freely given, partly because of his familiarity with military science dating from World War I. His will provided for an 80 acre arboretum near Princeton, symbolic of the outdoor life he so dearly loved.

Although men like Veblen gave counsel, American academic mathematicians as such were not at first called on to contribute to our war effort, perhaps because of their long neglect of applied mathematics. Indeed, Conant and Bush omitted mathematicians entirely from their original National Defense Research Council plans. However, by late 1942, Warren Weaver had obtained authorization for a mathematical analysis of the most effective setting for proximity fuzes in anti-aircraft shells, in spite of some opposition from Merle Tuve, their inventor. The committee making this analysis included Morse, von Neumann, H. H. Germond, and myself; I will never forget our initial 90 minute briefing by Weaver, which gave me my first mature insight into what applied mathematics was all about!

Morse, who had just served as AMS President, and was chairman of its War Preparedness Committee, continued throughout the war to make similar analyses of the optimal settings for proximity fuses in high explosive shells being fired at enemy ground troops, receiving a Meritorious Service Award from our army for this work and for his analysis of 'skip bombing.'

Von Neumann was active and brilliant both at Aberdeen and Los Alamos, where he worked with such later émigrés from Europe as Bethe, Fermi, and Teller, on the design of our first atomic bomb (see [29]). He early recognized the potentialities of general purpose digital computers, and his contributions to their design during the decade 1943–53 may well rank as his most important work (see [12]).

Wiener gained great prominence during the war through his imaginative ideas about cybernetics, of which one aspect was his stochastic prediction theory. Although his ideas stimulated the establishment of several research institutes and journals devoted to cybernetics, their technical impact on science has been limited so far.

After the war, Stone revamped the Mathematics Department at Chicago and became an international lecturer par excellence. His lecturing posts included Brazil, India, Japan, Turkey, and CERN in Geneva. He was president of the International Mathematical Union in 1952–4, and concluded his teaching career as George David Birkhoff Research Professor at the University of Massachusetts.

Walsh, faithful through the decades to the ideals and Weierstrassian analysis that had dominated Harvard mathematics when he entered Harvard as a sophomore around 1913, was rewarded by seeing approximation theory again become fashionable in the late 1960's, and by having the 'Walsh functions' which he had invented in the early 1920's admired by electrical engineers. Over 200 mathematicians came to a Symposium honoring his 75th birthday in 1970.‡ Walsh died a few years later, leaving over $200,000 for graduate scholarships in mathematics at Harvard.

I hope these brief concluding vignettes will give you some sense of how our once closely knit and patriotic mathematical community became internationalized, diversified, and even fragmented in the post-war years. To me, the most striking (and most poignant) mark of this change was the choice in 1950 of those two formidable Bourbaki exponents, Jean Dieudonné and André Weil, to write keynote articles in the *Monthly* about the nature of mathematics.**

If you will read these articles, then compare (or contrast!) them with those in the AMS Semicentennial Publications of 1938, and finally read Felix Klein's 1893 Evanston Colloquium Lectures and those by others at the 1893 Congress, you will get some sense of the extent to which the leaders I have described had achieved by 1938 their mission of matching Europe's mathematical culture, that their success had become an 'overkill' by 1941, and that it was taken for granted in a very changed world by 1950.

Today, American mathematicians are faced with different problems, at least as difficult as those which faced E. H. Moore, Osgood, Bôcher, Fine, Dickson, Bliss, Veblen, and G. D. Birkhoff in the 1890's and 1900's. Will they solve them as successfully? This is the challenge of the future!

Bibliography

1. R. C. Archibald, A Semicentennial History of the American Mathematical Society: 1888-1938, Semicentennial Publications, vol. I, Amer. Math. Soc., 1938.

2. Semicentennial Addresses of the American Mathematical Society, Semicentennial Publications, vol. II, Amer. Math. Soc., 1938.

3. E. T. Bell, Men of Mathematics, Simon and Schuster, New York, 1937.

4. ____, The Development of Mathematics, McGraw-Hill, New York, 1940.

5. George D. Birkhoff, Fifty Years of American Mathematics. (Same as [**2**, pp. 270-315], and [**6**, vol. iii, pp. 606-51].)

6. ____, Collected papers, 3 vols., Amer. Math. Soc., 1950; Dover reprint, 1968.

7. Garrett Birkhoff, The Rise of Modern Algebra, Men and Institutions in American Mathematics, Graduate Studies of Texas Tech. University, 1976, 41-85.

---

‡See vols. 5-6 of the Journal of Approximation Theory (1972), which give a good idea of Walsh's mathematical influence.

**N. Bourbaki, Amer. Math. Monthly 57 (1950) 221-32; A. Weil, ibid., 295-306.

8. ____, The William Lowell Putnam Mathematical Competition: Early History, Amer. Math. Monthly, 72 (1965) 469–83.

9. G. Birkhoff and Sue Ann Garwood, eds., The evolution of modern mathematics, Historia Math., 2 (1975) 425–624.

9a. M. Bôcher, D. R. Curtiss, P. F. Smith, E. B. van Vleck, Graduate work in mathematics . . . . . in the United States, Bull. Amer. Math. Soc., 18 (1911) 122–37.

10. J. L. Coolidge, Mathematics: 1870–1928, pp. 248–57 of S. E. Morison, The Development of Harvard University: 1869–1929, Harvard University Press, Cambridge, 1930. See also Coolidge's article in Amer. Math. Monthly, 50 (1943) 347–52.

11. T. C. Fry, Industrial mathematics, Amer. Math. Monthly, 48 (1941), special supplement (38 pp.), June–July, part II.

12. H. H. Goldstine, The Computer from Pascal to von Neumann, Princeton University Press, 1972. (Especially part I, ch. 9)

13. Morris Kline, Mathematical Thought from Ancient to Modern Times, Oxford University Press, New York, 1972.

14. K. O. May (ed.), The Mathematical Association of America: Its First Fifty Years, Mathematical Association of America, 1972.

15. J. von Neumann, Collected Works, six vols., Pergamon Press, New York, 1961, (A. H. Taub, editor).

16. John von Neumann: 1903–1957. Bull. Amer. Math. Soc., 64 (1958), no. 3, part 2 (129 pp.).

16a. Simon Newcomb, Reminiscences of an Astronomer, Harper, New York, 1903.

17. J. D. Tarwater, J. T. White, and J. D. Miller (eds.), Men and Institutions in American Mathematics, Graduate Studies Texas Tech., #13, 1976.

18. S. Timoshenko, History of Strength of Materials, McGraw-Hill, New York, 1953.

19. H. S. Vandiver, Some of my recollections of George D. Birkhoff, J. Math. Anal. Appl., 7 (1963) 27–83.

20. Warren Weaver, Scene of Change, Scribner, New York, 1970.

21. Hermann Weyl, Gesammelte Werke, four vols., Springer, New York, 1968. (K. Chandrasekharan, ed.). See also his Selecta.

22. Norbert Wiener, Ex-Prodigy, Simon and Schuster, New York, 1953.

23. _____, I am a Mathematician, Doubleday, Garden City, Long Island, 1956.

24. Norbert Wiener: 1894–1964, Bull. Amer. Math. Soc., 72 (1966) no. 1, part II.

25. Biographical Memoirs of the National Academy of Sciences (U.S.A.), published by the Academy.

26. R. C. Archibald, Opera Juvenum Americae . . . , Scripta Math., 4 (1937) 87–93, 188–195, 283–289, 330–334.

27. Dictionary of American Biography.

28. Dictionary of Scientific Biography, Charles C. Gillispie, ed., Scribner, New York, 1970.

29. S. M. Ulam, Adventures of a Mathematician, Scribner, New York, 1976.

30. Constance Reid, Courant, Scribner, New York, 1977.

AMERICAN MATHEMATICS FROM 1940 TO THE DAY BEFORE YESTERDAY*

J. H. Ewing, W. H. Gustafson, P. R. Halmos,
S. H. Moolgavkar, W. H. Wheeler, and W. P. Ziemer

1. Preface. What is the best way to present the small fragment of history described by the title above? Should this report occupy itself mainly with the statistics of the growth of Mathematical Reviews?, with the lives of mathematicians?, with lists of books and papers?, or with retracing the influences and implications that led from the bridges of Königsberg first to *analysis situs* and then to homological algebra? We decided to do none of these things, but, instead, to tell as much as possible about mathematics, the live mathematics of today. To do so within prescribed boundaries of time and space, we present the subject in the traditional "battles and kings" style of history. We try to describe some major victories of American mathematics since 1940, and mention the names of the winners, with, we hope, enough explanation (but just) to show who the enemy was. The descriptions usually get as far as statements only. We omit all proofs, but we sometimes give a brief sketch of how a proof might go. A sketch can be one sentence, or two or three paragraphs; its purpose is more to illuminate than to convince.

Progress in mathematics means the discovery of new concepts, new examples, new methods, or new facts. Schwartz's concept of distribution, Milnor's example of an exotic sphere, Cohen's method of forcing, and the Feit-Thompson theorem about simple groups are surely major by any standards. It was no trouble to find such victories to include in our list; the difficulty was to decide what to exclude. We formulated some rough rules

---

*The authors are grateful to W. Ambrose, G. Bennett, J. L. Doob, L. K. Durst, I. Kaplansky, R. Narashimhan, I. Reiner, and F. Treves for help, including advice, references, and, especially, encouragement.

This paper first appeared in the American Mathematical Monthly, 83 (1976) 503–516. Its preparation was supported in part by a grant from the National Science Foundation.

(e.g., theorems, not theories); since at least some aspects of applied mathematics were covered by other presentations, we restricted our attention to pure mathematics; we excluded work that had neither root, nor branch, nor flower in the U.S.; and, in deciding which of two candidates to keep, we leaned toward the one of greater general interest. ("Of general interest" is not quite the same as "famous", but it's close.)

We ended up with ten "battles and kings", and we think that they draw a fair picture of what's been happening. We do not say that our ten are greater than any others, nor that they are necessarily maximal in the mathematical sense of not being lesser than any others. We do say that they would all appear, and would be discussed with respect, in any responsible history of our place and time. The total number of such "non-omittable" victories is certainly greater than ten; it may be twenty or even forty. The choice of our ten was influenced by the limits of our competence and by our personal preferences; that could not be helped. Anyone else would very likely have selected a different set of ten. We hope and think, however, that everyone's list would have a large overlap with ours, and that the local differences would not essentially alter the global picture.

In history, every moment influences its successors; to restrict attention to a time interval may be often necessary, and sometimes possible, but it is rarely natural. In the same way, every place influences all others. Since the topology of the surface of our globe is much more intricate than that of the time line, to restrict attention to one country is almost impossible. The history of mathematics is no exception: trying to describe what happened *here*, we frequently yield to the pressure of distant influences and discuss what happened *there*. We were able to stay reasonably close to our original charge just the same; if fractional credits are assigned, something like 8.25 of the ten accomplishments described below can be called American. It might be of interest to observe also that over half of the original papers we refer to appeared in the *Annals of Mathematics*.

The order of the presentations might have been based on any of several principles (e.g., what actually happened first?, what is a prerequisite for what?). We decided to arrange them in order of complexity of the underlying category, or, in other words, very roughly speaking, in order of distance from the foundations of mathematics. At the end of each section there is a small list of pertinent references. The list is intentionally incomplete. All it contains is one (or, if necessary, two or three) of the earliest papers in which the discovery appears, and a more recent exposition of the discovery whenever we could find one.

2. Continuum Hypothesis. All mathematics is derived from set theory (or, in any event, many of us believe it is) and the manipulation of sets is a simple, natural exercise (or, in any event, students have very little trouble catching on to it). Everything that any working mathematician ever needs to know about sets (and a few extra things that he never thought he needed to know) could be summarized on one printed page (or three or four printed pages, if motivation is wanted along with the formalism). Such a page would state the basic ways of making new sets out of old (e.g., the formation of sets consisting of specified elements, the formation of unions of sets of sets, and the formation of the power set, i.e., the set of all subsets, of a set); it would describe the basic properties of sets (e.g., that two sets are equal if and only if each is a subset of the other, and that no set has elements that are themselves sets that have elements continued on downwards *ad infinitum*); and it would state (as an assumption or as a conclusion, but in either case as a description of the universe that sets live in) that infinite sets exist. These basic set-theoretic statements might be regarded either as obvious factual observations or as an axiomatic description of the ZF (Zermelo-Fraenkel) structure. In either case it would be a simple matter to code them in the language of a suitable (not very complicated) computer. Such a machine could easily be taught all the rules of inference that mathematicians ever use. If, in addition, its basic data were increased by two more statements, it could, in principle, easily print out all known mathematics (and a lot that is not yet known).

The two statements that history has subjected to extra scrutiny are AC (the axiom of choice) and GCH (the generalized continuum hypothesis). AC says that, for each set X, there is a function f from the power set of X into X itself such that $f(A) \in A$ for each non-empty subset A of X; GCH says that each subset of the power set of an infinite set X is in one-to-one correspondence either with some subset of X or with the entire power set — there is nothing in between.

Is AC true? The question has often been likened to a similar one about Euclid's parallel postulate. In both cases there is a more or less pleasant axiom system and a less pleasant, more complicated, non-obvious additional axiom. If the extra axiom is a consequence of the basic ones, it is true, and all is well; if its negation is a consequence of the basic ones, it is false, and, for better or for worse, the question is definitively answered. The same question can, of course, be asked about GCH. It has long been known that GCH implies AC; in view of this there is an obvious connection between the two answers.

The answers are subtle and profound intellectual achievements. Gödel proved (1940) that AC and GCH are not false (i.e., that they are consistent with the axioms of ZF), and Paul Cohen proved (1964) that they are not true (i.e., that they are independent of ZF).

Gödel argued by the construction of a suitable model. If, he said, ZF is consistent, so that there is a universe V of sets satisfying the basic axioms of ZF, then, he proved, there is a "sub-universe" that also satisfies them, and in which, moreover, both AC and GCH are true. The sub-universe Gödel constructed was the class L of "constructible" sets. (The word is given a very liberal but completely precise meaning; roughly speaking, the constructible sets are the ones that can be obtained from the empty set by a transfinite sequence of elementary set-theoretic constructions.) The class L is a substructure of V in the familiar mathematical sense of that word: the objects of L are some of the objects of V, and the relation $\in$ among them is the restriction of the set-theoretic $\in$ in V to the objects of L. The existence of a model such as L (constructed out of a hypothetically consistent model V) proves the consistency of AC and GCH the same way as the existence of the Euclidean plane proves the consistency of the parallel postulate.

Cohen's argument was similar but harder. It is reminiscent of Felix Klein's construction of a Lobachevskian plane by endowing a Euclidean disk with a new metric. Cohen started with a suitable model of ZF and adjoined new objects to it. The new objects are "classes" (but not sets) in the old model. The adjunctions proceed by a new method called "forcing", which, once it was discovered, was found to be applicable in many parts of set theory. Cohen's proof constructs an infinite sequence of better and better finite approximations to the new objects. Roughly speaking, each property of the new model is "forced" by properties of the old model and one of the approximations. Depending on how the details are adjusted, the end result can be a model of ZF in which AC is false, or a model of ZF in which AC is true but even the classical un-generalized continuum hypothesis CH is false. (CH is GCH for a countably infinite set.) Conclusion: AC and CH are independent of ZF.

References

1. P. J. Cohen, The independence of the continuum hypothesis, Proc. N. A. S., 50 (1963) 1143–1148 and 51 (1964) 105–110.

2. ____, Set theory and the continuum hypothesis, Benjamin, New York, 1966 (MR 38#999).

3. J. B. Rosser, Simplified independence proofs, Academic Press, New York, 1969 (MR 40#2536).

4. T. J. Jech, Lectures in set theory, with particular emphasis on the method of forcing, Springer, Berlin, 1971 (MR 48#105).

3. Diophantine Equations. The continuum hypothesis was the subject of Hilbert's first problem (in the famous list of 23 problems that he proposed in 1900); Hilbert's tenth problem concerned the solvability of Diophantine equations. The problem was to design an algorithm, a computational procedure, for determining whether an arbitrarily prescribed polynomial equation with integer coefficients has integer solutions. It is in some respects more natural and sometimes technically easier to discuss the *positive* integer solutions (solutions in $\mathbb{Z}_+$) of polynomial equations with *positive* integer coefficients. Caution: that does not mean equations such as $p(x) = 0$ only. The problem includes the search for x's such that $p(x) = q(x)$; more generally, it includes the search for n-tuples $(x_1, \ldots, x_n)$ such that $p(x_1, \ldots, x_n) = q(x_1, \ldots, x_n)$; and, in complete generality, it means the search for n-tuples $(x_1, \ldots, x_n)$ for which there exist m-tuples $(y_1, \ldots, y_m)$ such that

$$p(x_1, \ldots, x_n, y_1, \ldots, y_m) = q(x_1, \ldots, x_n, y_1, \ldots, y_m).$$

For each p and q (in $n + m$ variables) the solution set, in the latter sense, is called a "Diophantine set" in $\mathbb{Z}_+^n$.

What does it mean to say that there is an algorithm for deciding solvability? A reasonable way to answer the question is to offer a definition of computability for sets and functions, and then to define an algorithm in terms of computability.

When does a function from $\mathbb{Z}_+$ to $\mathbb{Z}_+$, or, more generally, a function from $\mathbb{Z}_+^n$ to $\mathbb{Z}_+$ deserve to be called "computable"? There is general agreement on the definition nowadays: computable functions (also called "recursive" functions) are the ones obtained from certain easy functions (constant, successor, coordinate) by three procedures (composition, minimalization, primitive recursion). The details do not matter here (they won't be used anyway); it might be comforting to know, however, that they are not at all difficult. A set (in $\mathbb{Z}_+$, or, more generally, in $\mathbb{Z}_+^n$) will be called computable in case its characteristic function is computable. Consequence: a set (in $\mathbb{Z}_+^n$) is computable if and only if its complement is computable.

Consider now all polynomial equations (in the sense described above), and let $\{E_1, E_2, E_3, \ldots\}$ be an enumeration of them. (In order for what follows to be in accord with the intuitive concept of an algorithm, the enumeration should be "effective" in some sense. That can be done, and it is relatively easy.) The indices k for which E_k has a solution (in the sense described above) form a subset S of $\mathbb{Z}_+$. The Hilbert problem (is there an algorithm?) can be expressed as follows: is S a computable set? The answer is no. The answer was a long time coming; it is the result of the cumulative efforts of J. Robinson (1952), M. Davis (1953), H. Putnam (1961), and Y. Matijasevič (1970).

The central concept in the proof is that of a Diophantine set, and the major step proves that every computable set is Diophantine. The techniques make ingenious use of elementary number theory (e.g., the Chinese remainder theorem, and a part of the theory of Fibonacci numbers, or, alternatively, of Pell's equation). The proof exhibits some interesting Diophantine sets whose Diophantine character is not at all obvious (e.g., the powers of 2, the factorials, and the primes).

One way to prove that S (the index set of the solvable equations) is not computable is by contradiction. If S were computable, then it would follow (by a slight bit of additional argument) that each particular Diophantine set (i.e., the solution set of each particular equation) is computable, and

hence (by the "major step" of the preceding paragraph) that the complement of every Diophantine set is Diophantine. The contradiction is derived
by exhibiting a Diophantine set whose complement is not Diophantine.

This last step uses a version of the familiar Cantor diagonal argument.
The idea is "effectively" to enumerate all Diophantine subsets of $\mathbb{Z}_+$, as
$\{D_1, D_2, D_3, \ldots\}$, say, prove that the set $D^* = \{n : n \in D_n\}$ is Diophantine
(that takes some argument), and, finally, to prove that the complement
$\mathbb{Z}_+ - D^* = \{n : n \notin D_n\}$ is not Diophantine (that's where Cantor comes in).

References

1. J. Robinson, Existential definability in arithmetic, Trans. Amer. Math. Soc., 72 (1952)
437–449 (MR 14-4).

2. M. Davis, Arithmetical problems and recursively enumerable predicates, J. Symbolic
Logic, 18 (1953) 33–41 (MR 14-1052).

3. M. Davis, H. Putnam, and J. Robinson, The decision problem for exponential
Diophantine equations, Ann. of Math., 74 (1961) 425–436 (MR 24#A3061).

4. Y. Matijasevič, The Diophantineness of enumerable sets (Russian), Dokl. Akad. Nauk
SSSR, 191 (1970) 279–282; improved English translation, Soviet Math. Dokl., 11 (1970)
354–358 (MR 41#3390).

5. M. Davis, Hilbert's tenth problem is unsolvable, Amer. Math. Monthly, 80 (1973)
233–269 (MR 47#6465).

4. Simple Groups. So much for the foundations. The next subject up
the ladder is algebra; in the present instance, group theory.

Every group G has two obvious normal subgroups, namely G itself and,
at the other extreme, the subgroup 1. A group is called "simple" if these
are all the normal subgroups it has.

Simple groups are like prime numbers in two ways: they have no proper
parts, and every finite group can be constructed out of them. (By general
agreement the trivial positive integer 1 is not called a prime, but the trivial
group 1 is called simple. Too bad, but that's how it is.)

Suppose, indeed, that G is finite, and let G_1 be a maximal normal subgroup of G. (To say that G_1 is maximal means that G_1 is a proper normal
subgroup of G that is not included in any other proper normal subgroup of
G.) If G is simple, then $G_1 = 1$; in any event, the maximality of G_1 implies
that the quotient group G/G_1 is simple. The relation between G, G_1, and
G/G_1 (group, normal subgroup, quotient group) is sometimes expressed by
saying that G is an extension of G/G_1 by G_1. In this language, every finite
group (except the trivial group 1) is an extension of a simple group by a
group of strictly smaller order. The statement is a group-theoretic analogue
of the number-theoretic one that says that every positive integer (except 1)
is the product of a prime by a strictly smaller positive integer.

If G_1 is not trivial, the preceding paragraph can be applied to it; the
result is a maximal normal subgroup G_2 in G_1, such that G_1 is an extension of the simple group G_1/G_2 by G_2. The procedure can be repeated so
long as it produces non-trivial subgroups; the end-product is a chain

$$G = G_0 \supset G_1 \supset G_2 \supset \cdots \supset G_n = 1$$

(a "composition series") with the property that each G_i/G_{i+1} is simple ($i = 0, \ldots, n - 1$). A great part of the problem of getting to know all finite groups reduces in this way to the determination of all finite simple groups. (The celebrated Jordan-Hölder-Schreier theorem is the comforting reassurance that, to within isomorphism, the composition factors G_i/G_{i+1} are uniquely determined by G, except for the order in which they occur.)

The abelian ones among the finite simple groups are easy to determine: they are just the cyclic groups of prime order. That's easy. What's hard is to find all non-abelian ones. Some examples of simple groups are easy to come by; among permutation groups, for instance, the most famous ones are the alternating groups of degree 5 or more. The known simple groups did not exhibit any pattern, and even the simplest questions about them resisted attack. Burnside conjectured, for instance, that every non-abelian simple group has even order, but that conjecture stood as an open problem for more than 50 years.

In a spectacular display of group-theoretic power, Feit and Thompson (1963) settled Burnside's conjecture (it is true). The proof occupies an entire issue (over 250 pages) of the *Pacific Journal.* It is technical group theory and character theory. Some reductions in it have been made since it appeared, but no short or easy proof has been discovered. The result has many consequences, and the methods also have been used to attack many other problems in the theory of finite groups; a subject that was once pronounced dead by many has shown itself capable of a vigorous new life.

Reference

1. W. Feit and J. G. Thompson, Solvability of groups of odd order, Pacific J. Math., 13 (1963) 775–1029 (MR 29#3538).

5. Resolution of Singularities. Algebra becomes richer, and harder, when it is mixed with and applied to geometry; one of the richest mixtures is the old but very vigorous subject known as algebraic geometry. This section reports the solution of an old and famous problem in that subject.

Let k be an algebraically closed field, and let k^n be, as usual, the n-dimensional coordinate space over k. (The heart of the matter in what follows will be visible to those who insist on sticking to the field of complex numbers in the role of k.) An "affine algebraic variety" V in k^n is the locus of common zeros of a collection of polynomials in n variables with coefficients in k. Since only the zeros matter, the collection itself is not important; it can be replaced by any other collection that yields the same locus. Thus, if R is the ring of *all* polynomials in n variables with coefficients in k, and if I is the ideal in R generated by the prescribed collection, then I will define the same variety; there is, therefore, no loss of generality in assuming that the collection was an ideal to begin with.

The objects of interest on varieties are their "singular points". Intuitively, these are points where the "tangent vectors" are not as they should be. Consider, for example, the curves defined by

$$y^2 = x^3 + x^2 \quad \text{and} \quad y^2 = x^3.$$

(Since the ground field was restricted to be algebraically closed, the *real* planar curves with these equations are not the right things to look at, but they are more lookable at than the complex curves, which lie in the complex plane. Warning: the complex plane has four real dimensions. To the algebraic geometer, the familiar "complex plane" of analysis is the complex *line*.) The first of these comes in to the origin from the first quadrant with slope 2, has a loop in the left half plane, and goes out from the origin to the fourth quadrant with slope -2; it has the origin as a double point. The other one comes in to the origin from the first quadrant with slope 0, and goes out the same way to the fourth quadrant; it has the origin as a cusp.

The effective way to deal with singular points begins by giving a purely algebraic description of them. Consider, for this purpose, the ring R_V of polynomial functions on V (i.e., the restrictions of the polynomials in R to V). If N_V is the ideal of R consisting of the polynomials that vanish on V, then, clearly, $R_V = R/N_V$. Each point $\alpha = (\alpha_1, \ldots, \alpha_n)$ of V induces a maximal ideal N_α in R (consisting of the set of polynomials that vanish at α); clearly $N_V \subset N_\alpha$.

The next step (in the program of defining singular points algebraically) is to form a new ring that studies the local behavior of functions near α. The idea is (very roughly) this. (i) Consider pairs (U, f), where U is a "neighborhood" of α and f is a rational function with no poles in U. (ii) Define an equivalence relation for pairs by writing $(U, f) \frown (U', f')$ exactly when there is a neighborhood U'' of α, included in $U \cap U'$, such that $f = f'$ on U''. (iii) The equivalence classes ("germs") form a ring (with, for example,

$$[(U, f)] + [(U', f')] = [(U \cap U', f + f')]),$$

called the "local ring" of V at α.

From the algebraic point of view, the preceding topological considerations are just heuristic; they will now be replaced by an algebraic construction. The process is, appropriately, called "localization". (i) Consider pairs (f, g), where f and g are in R and $g \notin N_\alpha$. (ii) Define an equivalence relation for pairs by writing $(f, g) \sim (f', g')$ exactly when there is an h not in $N\alpha$ such that $h \cdot (fg' - gf') = 0$. (iii) Write f/g for the equivalence class of (f, g). The equivalence classes form a ring R_α (with the usual rules of operations for fractions). The ring R_α is indeed a "local ring" in the customary algebraic sense: it has a *unique* maximal ideal, namely the one formed by the elements of R_α that vanish at α.

To motivate the next step, pretend, again, that the subject is not algebraic geometry, but analytic geometry. In that case R_α would consist of Taylor series at α convergent near α, and the ideal N_α of germs vanishing at α would consist of the Taylor series at α with vanishing constant term. The linear terms of a Taylor series are, in some sense, first order differentials. One way to capture just those terms is to "ignore" higher order terms. More precisely: consider the ideal N_α^2, which, in the analytic case, consists of the Taylor series with vanishing constant term and vanishing linear term, and form N_α/N_α^2.

The definition is now easy to formulate. The "dimension" d of V is, by definition, the minimum of the dimensions (over the field k, of course) of all the quotient spaces N_α/N_α^2; a point α is "singular" when dim $(N_\alpha/N_\alpha^2) >$ d. It is not difficult to see that for the two curves mentioned as examples above, the origin is indeed a singular point in the sense of this definition.

One of the main problems of algebraic geometry is to "get rid of" singular points. For this purpose the discussion is restricted to "irreducible" varities, i.e., to the ones for which R_V is an integral domain, or, equivalently, N_V is a prime ideal. In that case, form the field of fractions F_V of R_V. Two varieties V and W are "birationally equivalent" if F_V and F_W are isomorphic. This means roughly that V and W parametrize one another by rational mappings at all but finitely many places. The problem of "resolution of singularities" is that of finding a non-singular variety birationally equivalent to V.

The subject has a long history. Curves were handled by Max Noether in the 19th century. Surfaces were the subject of much geometric discussion by the Italian school; a rigorous proof was found by R. J. Walker (1935). For varieties of arbitrary dimension, over fields of characteristic 0, the final victory was inspired by Zariski's work; it was won by Hironaka (1964).

Reference

1. H. Hironaka, Resolution of singularities of an algebraic variety over a field of characteristic zero, Ann. of Math., 79 (1964) 109–326 (MR 33#7333).

6. Weil Conjectures.
The mathematician's work is often most difficult (and most rewarding) when he reasons by analogy, when he guesses that *this* situation ought to be just like *that* one. In 1949 A. Weil, reasoning in this way, proposed three conjectures that have profoundly influenced the development of algebraic geometry over the past 25 years.

The conjectures appeared in a paper entitled "Numbers of solutions of equations in finite fields", which was ostensibly a survey of previous work. Counting the number of solutions of a polynomial equation in several variables over a finite field was a classical problem, investigated by Gauss, Jacobi, Legendre, and others, but Weil took a new point of view. To understand his approach, consider the special case of the homogeneous equation

(†)
$$a_0 x_0^n + a_1 x_1^n + \cdots + a_r x_r^n = 0,$$

where the coefficients a_i are in the prime field F of p elements. The basic problem is to count the number of solutions in F, but, to number theorists, it is just as important to count the number of solutions in any finite extension field of F. Recall that, for every positive integer k, there is a unique extension field F_k of F with p^k elements. What Weil did was to count the number of solutions of (†) in each field F_k, and then code that information in a generating function.

To do this economically, examine the solution set of an equation such as (†). There is, of course, always the trivial solution, where all the x_i are zero; that one is justly regarded as trivial. If $(x_0, x_1, \ldots, x_r)$ is a non-trivial solution, and if $0 \neq c \in F_k$, then $(cx_0, cx_1, \ldots, cx_r)$ is also a non-trivial solution. Each non-trivial solution generates in this way $p^k - 1$ others, and there is no virtue in counting them all separately. It is natural, therefore, to consider the r-dimensional "projective space" $P^r(F_k)$, i.e., the set of non-trivial ordered $(r + 1)$-tuples of elements of F_k, where two are identified if one is a scalar multiple of the other. (This is exactly analogous to the familiar real and complex projective spaces.) The problem in these terms is to count the number of "points" in $P^r(F_k)$ that are "solutions" of (†).

That is precisely what Weil did. He let N_k be the number of solutions of (†) in $P^r(F_k)$, considered the generating function G,

$$G(u) = \sum_{k=1}^{\infty} N_k u^{k-1},$$

and proved a remarkable statement: G is the logarithmic derivative of a *rational* function. That is: there exists a rational function Z such that

$$\sum_{k=1}^{\infty} N_k u^{k-1} = \frac{d}{du} \log Z(u),$$

or, in other words, if

$$Z(u) = \exp \left(\sum_{k=1}^{\infty} \frac{N_k}{k} u^k \right),$$

then Z is rational. The function Z satisfies a functional equation analogous to the one satisfied by the Riemann zeta function, and it is appropriate to refer to Z as the zeta function associated with the equation (†). Motivated by classical problems that the Riemann zeta function gave rise to, Weil studied and was able to determine many properties of the zeros and the poles of Z.

Here is where Weil's paper reaches its climax. Weil wanted to extend the results about (†) to algebraic varieties in $P^r(F_k)$, i.e., to the solution sets of *systems* of homogeneous equations in r variables. The notion of a zeta function, originally defined by Riemann, was extended by Dedekind to algebraic number fields, by Artin to function fields, and now, by Weil, to algebraic varieties. (The varieties to be considered should be non-singular. It doesn't matter here what the general definition of that condition is; for most fields it can be defined as usual by requiring that the Jacobian of the system of equations have maximal rank at every point.) Given a system of equations, with coefficients in F, let N_k be, as before, the number of solutions in $P^r(F_k)$. Weil advanced the following conjectures. One: the function Z, defined as before by

$$Z(u) = \exp \left(\sum_{k=1}^{\infty} \frac{N_k}{k} u^k \right),$$

is rational. Two: Z satisfies a particular functional equation, which, as before, bears a striking resemblance to the one satisfied by the Riemann zeta function. Three: the reciprocals of the zeros and the poles of Z are algebraic integers and their absolute values are powers of $\sqrt{p}$. (This is called the generalized Riemann hypothesis.)

All this might seem far removed from what is normally thought of as geometry, and, although several examples were known, it might seem that Weil made his conjectures on strikingly little evidence. What was really behind the conjectures? The answer is contained in the last paragraph of Weil's paper, where he suggests that there is an analogy between the behavior of these varieties (for fields of characteristic p) and that of the classical varieties (for the field of complex numbers).

In 1960 Dwork established the rationality conjecture (without the condition of non-singularity). The final triumph came in 1974: using twenty years' of results of the Grothendieck school, Deligne established all the Weil conjectures, and, perhaps more importantly, proved that there is a beautiful connection between the theory of varieties over fields of characteristic p and classical algebraic geometry. "God ever geometrizes", said Plato, and "God ever arithmetizes", said Jacobi; the Weil conjectures show, better than anything else, how He can do both at once.

References

1. A. Weil, Numbers of solutions of equations in finite fields, Bull. Amer. Math. Soc., 55 (1949) 497–508 (MR 10-592).

2. P. Deligne, La conjecture de Weil I, Inst. Hautes Études Sci. Publ. Math., 43 (1974) 273–307 (MR 49#5013).

3. J. A. Dieudonné, The Weil conjectures, The Mathematical Intelligencer, 10 (September 1975) 7–21.

7. Lie Groups. So much for algebra, with or without geometry. The next subject points toward some of the later analytic ones by mixing algebra with topology. The result, like a few other outstanding results of mathematics, seems to get something for nothing, or, at the very least, to get quite a lot for an astonishingly low price. One of the most famous results of this kind occurs in the early part of courses on complex function theory: it asserts that a differentiable function on an open subset of the complex plane is necessarily analytic.

Hilbert's fifth problem asked for such a something-for-nothing result. The context is the theory of topological groups. A topological group is a set that is both a Hausdorff space and a group, in such a way that the group operations

$$(x, y) \mapsto xy \qquad \text{and} \qquad x \mapsto x^{-1}$$

are continuous. A typical example is the set of all 2×2 real matrices of the form $\left(\begin{smallmatrix} x & y \\ 0 & 1 \end{smallmatrix}\right)$ with $x > 0$; the topological structure is that of the right half plane (all (x, y) with $x > 0$), and the multiplicative structure is the usual one associated with matrices. Equivalently: define multiplication in the right half plane by

$$(x, y) \cdot (x', y') = (xx', xy' + y);$$

since

$$(x, y)^{-1} = \left(\frac{1}{x}, \frac{-y}{x} \right),$$

it is clear that both multiplication and inversion are continuous.

This example has an important special property: it is "locally Euclidean" in the sense that every point has a neighborhood that is homeomorphic to an open ball in (2-dimensional) Euclidean space. (Equivalently: every point has a "local coordinate system".) An even more important special property of the example is that the group operations, regarded as functions on the appropriate Euclidean space, are not only continuous but even analytic. If a group is locally Euclidean, i.e., if it can be "coordinatized" at all, then there are many ways of coordinatizing it; if at least one of them is such that the group operations are analytic, the group is called a "Lie group". Hilbert's fifth problem was this: is every locally Euclidean group a Lie group?

The analogy of this problem with the one in complex function theory is quite close. It is relatively elementary that a twice-differentiable function is analytic; it has been known for a long time that if a topological group has sufficiently differentiable coordinates, then it has analytic ones.

Immediately after the discovery of Haar measure, von Neumann (1933) applied it to prove that the answer to Hilbert's question is yes for compact groups. A little later Pontrjagin (1939) solved the abelian case, and Chevalley (1941) solved the solvable case. (Sorry about that, but "solvable" is a technical word here and its use is unavoidable.)

The general case was solved in 1952 by Gleason and, jointly, by Montgomery and Zippin; the answer to Hilbert's question is yes. What Gleason did was to characterize Lie groups. (Definition: a topological group "has no small subgroups" if it has a neighborhood of the identity that includes no subgroups of order greater than 1. Characterization: a finite-dimensional locally compact group with no small subgroups is a Lie group.) Montgomery and Zippin used geometric-topological tools (and Gleason's theorem) to reach the desired conclusion.

Warning: the subject cannot be considered closed. The question can be generalized in ways that are both theoretically and practically valuable. Groups can be replaced by "local groups", and abstract groups can be replaced by groups of transformations acting on manifolds. The best kind of victory is the kind that indicates where to look for new worlds to conquer, and the one over Hilbert's fifth problem was that kind.

References

1. A. Gleason, Groups without small subgroups, Ann. of Math., 56 (1952) 193–212 (MR 14–135).

2. D. Montgomery and L. Zippin, Small subgroups of finite-dimensional groups, Ann. of Math., 56 (1952) 213–241 (MR 14–135).

3. ____,Topological transformation groups, Interscience, New York, 1955 (MR 17–383).

8. Poincaré Conjecture. A "manifold" is a topological space (a separable Hausdorff space to be exact) that is locally Euclidean. Manifolds have been the central subject of topology for many years, and still are. Hilbert's fifth problem was about group manifolds; the Poincaré conjecture is about the connectedness properties of smooth manifolds. A "differential manifold" is a manifold endowed with local coordinate systems such that the change of coordinates from one coordinate neighborhood to an overlapping one is smooth. "Smooth" in this context is a generally accepted abbreviation for C^∞, i.e., for infinitely differentiable.

The axioms of Euclidean plane geometry characterize the plane. This kind of activity (find the central core of a subject, abstract it, and use the result as an axiomatic characterization) is frequent and useful in mathematics. Since spheres are the principal concept of a large part of topology, it is natural to try to subject them too to the axiomatic approach. The attempt has been made, and, to a large extent, it was successful.

The 1-sphere, for instance (i.e., the circle), is a compact, connected 1-manifold (i.e., a manifold of dimension 1), and that's all it is: to within a homeomorphism every compact connected 1-manifold is a 1-sphere.

For the 2-sphere, the facts are more complicated: both the 2-sphere S^2 and the torus T^2 ($= S^1 \times S^1$) are compact connected 2-manifolds, and they are not homeomorphic to each other. To distinguish S^2 from T^2 and, more generally, from a sphere with many handles, it is necessary to observe that, although both S^2 and T^2 are connected, S^2 is more connected. In the appropriate technical language, S^2 is "simply connected" and T^2 is not. The relevant definitions go as follows. Suppose that X and Y are topological spaces and that f and g are continuous functions from X to Y; write I for the unit interval $[0, 1]$. The functions f and g are "homotopic" if there exists a continuous function h from $X \times I$ to Y such that $h(x, 0) = f(x)$ and $h(x, 1) = g(x)$ for all x. (Intuitively: f can be continuously deformed to g.) The space Y is simply connected if every continuous function from S^1 to Y is homotopic to a constant. (Intuitively: every closed curve can be shrunk to a point.) Once this concept is at hand, the characterization of the 2-sphere becomes easy to state: to within a homeomorphism, every compact, connected, simply connected 2-manifold is a 2-sphere.

The discussion of dimensions 1 and 2 does not yet provide a firm basis for guessing the general case, but it does at least make the following concept plausible. There is a way of defining "k-connected" that generalizes "connected" ($k = 0$) and "simply connected" ($k = 1$): just replace S^1 in the definition of simple connectivity by S^j, $j = 0, 1, \ldots, k$. Thus: a space Y is k-connected if, for each j between 0 and k inclusive, every continuous function from S^j to Y is homotopic to a constant.

The general Poincaré conjecture is that if a smooth compact n-manifold is $(n - 1)$-connected, then it is homeomorphic to S^n. For $n = 1$ and $n = 2$ the result has been known for a long time; the big recent step was the proof of the assertion for all $n \geqslant 5$. The proof was obtained by Smale (1960). Shortly thereafter, having heard of Smale's success, Stallings gave another proof for $n \geqslant 7$ (1960) and Zeeman extended it to $n = 5$ and $n = 6$ (1961). For $n = 3$ (the original Poincaré conjecture) and for $n = 4$ the facts are not yet known.

Actually Smale proved a much stronger result. He showed how certain manifolds could be obtained by gluing disks together. His results provide a starting point for a classification of simply connected manifolds.

References

1. S. Smale, The generalized Poincaré conjecture in higher dimensions, Bull. Amer. Math. Soc., 66 (1960) 373–375 (MR 23#A2220).

2. J. R. Stallings, Polyhedral homotopy-spheres, Bull. Amer. Math. Soc., 66 (1960) 485–488 (MR 23#A2214).

3. E. C. Zeeman, The generalized Poincaré conjecture, Bull. Amer. Math. Soc., 67 (1961) 270 (MR 23#A2215).

4. S. Smale, Generalized Poincaré's conjecture in dimensions greater than four, Ann. of Math., 74 (1961) 391–406 (MR 25#580).

9. Exotic Spheres.

A "diffeomorphism" between two differential manifolds is a homeomorphism such that both it and its inverse are smooth. Homeomorphism is an equivalence relation between manifolds; the equivalence classes (homeomorphism classes) consist of manifolds with the same topological properties. Similarly, diffeomorphism is an equivalence relation between differential manifolds, and the equivalence classes (diffeomorphism classes) consist of manifolds with the same differential properties. Are these concepts really different? Is diffeomorphism really more stringent than homeomorphism? The answer is yes, even for topologically very well-behaved manifolds, but that is far from obvious. An example constructed by Milnor in 1956 came as a surprise, and, according to Hassler Whitney, that single, isolated example led to the modern flowering of differential topology.

Milnor's example is the 7-sphere. For every positive integer n, the n-sphere S^n is embedded in Euclidean $(n + 1)$-space in a natural way, and thus has a natural differential structure. Milnor showed that there exists a differential manifold that is homeomorphic but not diffeomorphic to S^7; such a manifold has come to be called an "exotic" 7-sphere.

To prove the assertion, there are three problems to solve: (1) find a candidate, (2) prove that it is homeomorphic to S^7, and (3) prove that it is not diffeomorphic to S^7. The first problem was easy (with hindsight); the candidate was a space (a 3-sphere bundle over the 4-sphere) that had been familiar to topologists for a number of years. Milnor solved the second problem using Morse theory. A Morse function on a differential manifold is a real-valued smooth function with only non-degenerate critical points. The n-sphere has a Morse function with exactly two critical points (project onto the last coordinate and consider two poles). A theorem of G. Reeb's goes in the other direction: if a differential manifold has a Morse function with exactly two critical points, then it is homeomorphic to a sphere. Milnor showed that his candidate had such a Morse function. The third problem was the hardest. Here Milnor used two facts: first, that S^7 is the boundary of the unit ball in $\mathbb{R}^8$, and, second, that his candidate was presented as the boundary of an 8-dimensional manifold W. If the candidate were diffeomorphic to S^7, then, using the diffeomorphism, one could glue the unit ball onto W and obtain an 8-dimensional manifold that (as Milnor showed) cannot exist.

Once it was known that exotic 7-spheres could exist, it was natural to ask how many there were, i.e., how many diffeomorphism classes there were. Milnor and Kervaire showed that there are 28. What about the other spheres? Again Milnor and Kervaire showed that the set of differential n-spheres (modulo diffeomorphism) could be made into a finite abelian group, with the "natural" sphere as the zero element; the group operation is the "connected sum", which is the natural gluing together of manifolds. The group is trivial for $n < 7$; it has order 28 for $n = 7$, order 2 for $n = 8$, order 8 for $n = 9$, order 6 for $n = 10$, and order 992 for $n = 11$. For $n = 31$, there are over sixteen million (diffeomorphism classes of) exotic spheres.

There are two systematic ways of constructing exotic spheres. The first is Milnor's "plumbing" construction (joining holes by tubes), which presents exotic spheres as boundaries of manifolds assembled by cutting and pasting. The other method (due to Brieskorn, Pham, and others) gives preassembled examples. For each finite sequence $(a_1, \ldots, a_n)$ of positive integers, let $\Sigma (a_1, \ldots, a_n)$ be the set of those zeros of the polynomial $z_1^{a_1} + \cdots + z_n^{a_n}$ that lie on the unit sphere in complex n-space. Milnor gave precise criteria on the n-tuple that ensure that this manifold is homeomorphic to a sphere of the appropriate dimension (which is $2n - 3$, by the way). For example, as k runs from 1 to 28, the manifolds $\Sigma (3, 6k - 1, 2, 2, 2)$ provide representatives for the 28 different diffeomorphism classes of 7 spheres.

References

1. J. W. Milnor, On manifolds homeomorphic to the 7-sphere, Ann. of Math., 64 (1956) 399–405 (MR 18-498).

2. ____, Differential Topology, Lectures on modern mathematics, vol. II, pp. 165–183, Wiley, New York, 1964 (MR 31#2731).

10. Differential Equations. Differential concepts play an important role everywhere, including pure algebra and, as above, topology. Differential equations are what make the world go around, and anyone who wants to predict and perhaps partly to change how the world goes around must know about differential equations and their solutions.

Differential equations are classified in a curiously primitive manner according to the number of independent variables that are involved in differentiation, and the way in which the unknown functions enter. The classification is "one" and "many" in the one case, and "good" and "not-so good" in the other, or, in terms of the corresponding adjectives that apply to the equations, "ordinary" and "partial" in the one case, and "linear" and "non-linear" in the other. This report is concerned with linear equations only, and partial ones at that; ordinary ones make just a brief appearance at the beginning, to set the stage.

The beginnings of the theory of ordinary linear differential equations are simple and satisfactory; they can be found in elementary textbooks. If p is a polynomial

$$p(\xi) = \sum_{j=0}^{k} a_j \xi^j,$$

and if $D = d/dx$, then $P = p(D)$ is a differential operator, and $Pu = g$ (for given g and unknown u) is the typical linear O.D.E. with constant coefficients. If g is continuous (a reasonable, useful, but much too special assumption), then the equation always has a solution. The conclusion remains true even for variable coefficients (i.e., in case the a_j's themselves are functions of x), provided they are subjected to some mild restrictions. It is, for instance, sufficient that the a_j's be continuous and that the "principal" coefficient a_k have no zeros.

For partial differential equations even the beginnings are non-trivial and new, and, for instance, even the theory for constant coefficients belongs to the most recent period of research. The formulation of the problem is easy enough: consider a polynomial in several variables $\xi_1, \ldots, \xi_n$, and obtain a differential operator P by replacing ξ_j by $\partial / \partial x_j$; the problem is to solve $Pu = g$ for u.

To avoid some not especially enlightening and not especially useful epsilontic hairsplitting, it has become customary to take g (and to seek u) in either the most or the least restrictive class of objects in sight. The most restrictive class consists of the smooth (infinitely differentiable) functions on whatever domain is under consideration ($\mathbb{R}^n$, an open set in $\mathbb{R}^n$, a manifold); the other extreme is represented by Laurent Schwartz's distributions. (The motivation of distribution theory is that functions f induce linear functionals $\phi \mapsto \int \phi (x) f(x) dx$ on C^∞. A "distribution" is a suitably continuous linear functional, not necessarily one induced by a function. The analogy between the generalization and its source suggests an appropriate definition of differentiation for distributions, and with that definition the theory of partial differential equations is off and running.)

Partial differential equations is an old subject and a widely applied one, and it is astonishing that the basic theorem is as recent as it is; it seems only the day before yesterday that Ehrenpreis (1954) and Malgrange (1955) proved that every linear P.D.E. with constant coefficients is solvable. If the right hand side is smooth, there is a smooth solution; even if the right hand side is allowed to be an arbitrary distribution, there is a distribution solution. The subject is exhaustively treated in Ehrenpreis's book (1962) and can be regarded as closed.

So far so good; the proofs are harder than for O.D.E.'s, but the facts are pleasant. The theory for variable (i.e., function) coefficients is much harder, much less known, and nowhere near finished. Two exciting contributions to it in the late 1950's showed that old guesses and old methods were woefully inadequate.

As for old guesses: Hans Lewy produced (1957) an inspired and amazingly simple example of a P.D.E. with variable (but *very* smooth) coefficients that has no solutions at all. Lewy's polynomial is of degree 1,

$$p(x, \xi) = a_1 \xi_1 + a_2 \xi_2 + a_3 \xi_3,$$

where the coefficients a_1, a_2, a_3 are functions of three variables x_1, x_2, x_3, and, in fact, the first two are constants:

$$a_1 = -i, \quad a_2 = 1, \quad a_3 = -2(x_1 + ix_2).$$

The corresponding differential operator is, of course,

$$P = -i\frac{\partial}{\partial x_1} + \frac{\partial}{\partial x_2} - 2(x_1 + ix_2)\frac{\partial}{\partial x_3}.$$

What Lewy proved is that for almost every g in C^∞ (in the sense of Baire category) the equation $Pu = g$ is satisfied by no distribution whatever.

At about the same time (1958) Calderón studied the uniqueness of the solution of certain important partial differential equations (under suitable initial conditions). He showed, in effect, that if $Pu = 0$, with $u = 0$ for $t \leqslant 0$ (intuitively, "t" here is time), then, locally u remains 0 for some positive time. Calderón's methods were transplanted from harmonic analysis; they introduced singular integrals into the subject, whence, a little later, came pseudo-differential operators and Fourier integral operators. These ideas have dominated the subject ever since.

Hörmander analyzed and generalized Lewy's example (1960). What makes it work, he pointed out, was that the coefficients are complex; what is fundamental is the behavior of the commutator of P and $\bar{P}$. The operator $\bar{P}$ here is obtained simply by replacing each coefficient by its complex conjugate. (In operator language: $\bar{P}u = \overline{(P\bar{u})}$.) More precisely: consider, for each polynomial in $(\xi_1, \ldots, \xi_n)$ its "principal part", i.e., the part that involves the terms of highest degree only. (For Lewy's example there is no other part.) If $p(x, \xi)$ is the principal part, write $b(x, \xi)$ for the "Poisson bracket",

$$b(x, \xi) = \sum_j \left(\frac{\partial p}{\partial \xi_j} \frac{\partial \bar{p}}{\partial x_j} - \frac{\partial p}{\partial x_j} \frac{\partial \bar{p}}{\partial \xi_j} \right).$$

Assertion: if, for some (x^0, ξ^0), the principal part $p(x^0, \xi^0)$ vanishes but the Poisson bracket $b(x^0, \xi^0)$ does not, then p is, in the sense of Lewy, not solvable in any open set containing x^0. It is easy to see that the Lewy example is covered by the Hörmander umbrella. Indeed: since

$$p = -i\xi_1 + \xi_2 - 2(x_1 + ix_2)\xi_3,$$

$$\bar{p} = i\xi_1 + \xi_2 - 2(x_1 - ix_2)\xi_3,$$

elementary computation yields

$$b = 8i\xi_3,$$

and it becomes clear that for every $x = (x_1, x_2, x_3)$ there is a $\xi = (\xi_1, \xi_2, \xi_3)$ such that $p(x, \xi) = 0$ and $b(x, \xi) \neq 0$.

References

1. L. Ehrenpreis, Solution of some problems of division, Amer. J. Math., 76 (1954) 883–903 (MR 16-834).

2. B. Malgrange, Existence et approximation des solutions des équations aux dérivées partielles et des équations de convolution, Ann. Inst. Fourier, Grenoble, 6 (1955) 271–355 (MR 19-280).

3. H. Lewy, An example of a smooth linear partial differential equation without solution, Ann. of Math., 66 (1957) 155–158 (MR 19-551).

4. A. P. Calderón, Uniqueness in the Cauchy problem for partial differential equations, Amer. J. Math., 80 (1958) 16–36 (MR 21#3675).

5. L. Hörmander, Differential operators of principal type, Math. Ann., 140 (1960) 124–146 (MR 24#A434).

6. ____, Linear partial differential equations, Springer, New York, 1969 (MR 40#1687).

7. L. Ehrenpreis, Fourier analysis in several complex variables, Wiley, New York, 1970 (MR 44#3066).

11. Index Theorem. The Atiyah-Singer index theorem (1963) spans two areas of mathematics, topology and analysis, and that's not an accident of technique but in the nature of the subject: the span is what it's all about. Theorems with such a broad perspective are usually the ones that are the most useful and the most elegant, and the index theorem is no exception. The very breadth of the theorem requires, however, that an expository sketch of it proceed obliquely. In what follows we describe, first and mainly, a historical and conceptual precursor, the Riemann-Roch theorem, and then indicate, briefly, how the Atiyah-Singer theorem generalizes it.

The classical Riemann-Roch theorem deals with the dual nature (topological and analytic) of a Riemann surface. Every compact Riemann surface is homeomorphic to a (two-dimensional) sphere with handles. The number of handles, the "genus", completely determines the topological character of the surface; that part is easy. The analytic structure is more complicated. It consists of a covering by a finite number of open sets and of explicit homeomorphisms from the complex plane $\mathbb{C}$ to each open set, which define holomorphic functions on the overlaps. (It is convenient and harmless to use the homeomorphisms to identify each open set in the covering with an open set in $\mathbb{C}$; that is tacitly done below.) If, for example, the surface is the sphere (with no handles), think of $\mathbb{C}$ as slicing through the equator, and use stereographic projections (toward the north and south poles) as the homeomorphisms. There are two open sets here, the complement of the north pole and the complement of the south pole; the holomorphic function of the overlap is given by $w(z) = 1/z$.

A smooth function on a Riemann surface can be viewed as a set of functions on, say, the open unit disk $\mathbb{C}$ (one for each of the open sets of the covering) that are smooth (C^∞) and transform into one another under the changes of variables induced by the overlaps. (If, that is, f and g are two of these functions, and w is the transformation on the disk induced by going via the appropriate homeomorphism to the open set corresponding to f and coming back from the overlap with the open set corresponding to g, then $f(z) = g(w(z))$.) The function on the Riemann surface is called holomorphic (or meromorphic) if each of these functions on the disk is holomorphic (or meromorphic). Another necessary concept for the analytic study of a Riemann surface are constants: that is essentially what Liouville's theorem the form $p(x, y)dx + q(x, y)dy$, where p and q are complex-valued smooth functions that, on the overlaps, satisfy the chain rule for change of variables. A holomorphic differential is one of the form $f(z)dz$, where f is holomorphic and $dz = dx + i\,dy$. (In the notation used above, the overlap relation for these differentials becomes $f(z)dz = g(w)dw = g(w(z))w'(z)dz$; the functions f and g no longer merely transform into one another, but are altered by the contribution of the differentials as well.)

The analytic properties of a Riemann surface are the properties of the holomorphic (and meromorphic) functions and differentials that it possesses. A well-known result is that the only holomorphic functions on a compact Riemann surface are constants: that is essentially what Liouville's theorem says. The Riemann-Roch theorem says much more. In its simplest form it deals with a compact Riemann surface S of genus g, and n points $z_1, \ldots, z_n$ on S. Let F be the vector space of meromorphic functions on S with poles of order not greater than 1 at each z_i (and nowhere else); let D be the vector space of holomorphic differentials with zeros of order not less than 1 at each z_i (and possibly elsewhere). Conclusion:

$$\dim F - \dim D = 1 + n - g.$$

(In the special case of the classical Liouville theorem, $g = 0$, $n = 0$, and $\dim D = 0$.) The important aspect of the conclusion is that a quantity described completely in *analytic* terms can be computed from nothing but *topological* data.

In the special case $n = 0$, F is the vector space of holomorphic functions on S (so that $\dim F = 1$) and D is the space of all holomorphic differentials. There is a linear map, conventionally denoted by $\bar{\partial}$, from the vector space of all smooth functions on S to the vector space of all smooth differentials: write

$$\bar{\partial}f = \frac{\partial f}{\partial \bar{z}}\,d\bar{z}$$

in each of the open sets of the prescribed covering. The map $\bar{\partial}$ is an example of a differential operator. The kernel of $\bar{\partial}$ consists precisely of the functions satisfying the Cauchy-Riemann equations; in other words

$$\ker \bar{\partial} = F.$$

The cokernel of $\bar{\partial}$ (the quotient space of the space of all smooth differentials modulo the image of $\bar{\partial}$) is similarly identifiable with D. The conclusion of the Riemann-Roch theorem takes, in this case, the form

$$\dim \ker \bar{\partial} - \dim \operatorname{coker} \bar{\partial} = 1 - g.$$

The Atiyah-Singer theorem is a generalization of the Riemann-Roch theorem in that it too states that a certain analytically defined number (the "analytic index") can be computed in terms of topological data. Which aspects are generalized? All. To begin with, the Riemann surface is replaced by an arbitrary compact smooth manifold M of arbitrary dimension. The vector spaces of smooth functions and smooth differentials are replaced by vector spaces of smooth sections of complex vector bundles over M (in fact, complexes of vector bundles). The map $\bar{\partial}$, finally, is replaced by a differential operator Δ, which satisfies a certain invertibility condition (called ellipticity). It follows that both $\ker \Delta$ and $\operatorname{coker} \Delta$ are finite-dimensional; the difference of the two dimensions is the analytic index. The conclusion is that the analytic index can be computed in terms of topological invariants (the "topological index"), which are very sophisticated generalizations of the genus.

Even in its relatively short life the Atiyah-Singer index theorem has had important and interesting consequences, and has been proved in at least three enlighteningly different ways. A recent proof depends on the study of the heat equation on a manifold.

References

1. M. F. Atiyah and I. M. Singer, The index of elliptic operators on compact manifolds, Bull. Amer. Math. Soc., 69 (1963) 422–433 (MR 28#626).

2. R. S. Palais, Seminar on the Atiyah-Singer index theorem, Ann. Math. Studies, No. 57, Princeton University Press, Princeton, 1965 (MR 33#6649).

12. Epilogue. Concepts, examples, methods, and facts continue to be discovered; problems get reformulated, placed in new contexts, better understood, and solved every day. We hope that the ten examples above have communicated at least a part of the breadth, depth, excitement, and power of the mathematics of our time. Mathematics is alive, and it's here to stay.

MATHEMATICS AND THE GOVERNMENT: THE POST-WAR YEARS AS AUGURY OF THE FUTURE

Mina S. Rees

My topic is Mathematics and the Government; I shall be addressing myself to the years following World War II, years during which significant support of scientific research by the Federal Government stimulated an expansion of scientific activity that placed the United States in the forefront of scientific achievement. The level and scope of this federal support was a new phenomenon, a fall-out from the participation of scientists and mathematicians in the scientific work that aided the country's military efforts in World War II. Though mathematicians had certainly been involved in government programs long before the war, e.g., in the work of the Coast and Geodetic survey, and of the Agricultural Extension Service, the substantial support of the research of university mathematicians that followed the war introduced a new dimension into the relationships between the government and mathematicians, and changed the life style of many university people. I shall try to record some aspects of the government's activities and of mathematicians' responses to government overtures from the time of the establishment of the Office of Naval Research immediately after the war till about the time of the establishment of the National Science Foundation in 1950. I shall record some of the ways in which government-supported research resulted in the flowering of new or previously neglected fields which, since the time of their revival, have blossomed and become part of the familiar mathematical scene on college and university campuses.

For many of our younger mathematicians, the idea of a system of graduate education with no government fellowships, no grant support from government agencies, no government contracts must seem strange indeed. Yet all these things were almost unheard of in mathematics departments before World War II. Fellowship support, when it was available to mathematicians, was usually provided by the university itself or funded by one of the private foundations; and contract support for university research was largely initiated by the Navy with the establishment of the Office of Naval Research in 1946. It should, perhaps, be emphasized that that office was never authorized to support educational programs and therefore never established a program of fellowships. The device that was accepted in lieu of fellowships was originally the research assistantship which has earned an honorable place in the continuing programs of the National Science Foundation and other governmental agencies, and later the Research Associateship, which provided post-doctoral support.

For many of the older mathematicians who received their degrees in the 1930's or early 1940's, my topic must conjure up not only a picture of government support of research and education but also the memory of a variety of assignments during World War II. For those who found themselves in uniform and who were very lucky, the job was to serve as a mathematician in an Army or Navy installation that needed mathematicians. One conspicuous example of this was Herman Goldstine's assignment to the Army's Aberdeen Proving Ground—an assignment that had a profound effect upon his subsequent career. For others there were posts in industry or universities working on wartime undertakings that had mathematical components; and for still others there was work in Operations Research Groups attached to Army units abroad. But the activity that had by far the greatest influence on the post-war development of mathematics was the Office of Scientific Research and Development (OSRD) through the operation of its Applied Mathematics Panel. The OSRD adapted the concept of the procurement contract routinely used by the government for the purchase of goods and equipment, to embrace the purchase from the universities of research and development in support of military needs. By encouraging a limited number of universities to bring together research mathematicians to work under such contracts, the Applied Mathematics Panel made mathematical assistance available to large numbers of natural scientists and engineers who provided scientific support for the war effort. The wartime performance of these scientists won for science the high regard of the military establishment and of Congress and the recognition that post-war expansion of research in the sciences was a national requirement. This attitude reflected the conviction that the sciences must be strong if the country was to maintain itself in the competition for military security, industrial expansion, and the material well-being of its people which, in those days, every one was sure would flow from the applications of scientific results to technology.

The importance of a strong base of science and technology for the welfare of an industrial society is still generally acknowledged. In a UNESCO report on *National Science Policies of the U.S.A.* published in 1968, the point is emphasized: [1]

There is a general consensus that science and technology are indispensable to the economic growth of an industrially advanced society ... An acceptable theory of the economic value of research and development is not yet at hand ... [but] .. empirical studies of R&D indicate that there is a general strong correlation between quantity of R&D investment and the resulting number of innovations ... An issue of growing national concern has been the distribution of scientific capabilities and of federal support by geographical regions. This relates to two important national objectives: the provision of maximum educational opportunity, according only to ability and motivation, to all segments of the population, and assurance of equal opportunity for regional economic development .... On the other hand, much R&D effort is specifically oriented to non-economic aspects of the public interest and is not supported for purposes of its contribution to the economy in the normal sense of the term ... funds are also directly allocated to basic science in recognition of its role as an essential underpinning for the general social, political and economic objectives of the nation ... much basic research has been supported by the different agencies of government and by private industry in areas considered relevant to their interests. Because these interests can fluctuate, and sometimes become narrow, the National Science Foundation has responsibility for maintaining "balance" and the overall "health" of basic research, particularly in connection with higher education ... It is generally conceded that the federal support which has made possible so much of the academic research is of great benefit. But there are also criticisms of the effects such support has had.

During World War II, however, and immediately after the war, criticisms and doubts were muted, if they were present at all. Because many of the criticisms on university campuses during the late 1960's were directed toward the considerable reliance of universities on military support of basic research, I believe it is worth quoting a description of the World War II atmosphere as related by Harvey Brooks:[2]

In a certain sense World War II and the subsequent period of the Cold War might be characterized as a love affair between the intellectual community and the government, which affected not only the development of science but a much broader range of academic scholarship. The Nazi menace united the American intellectual community as nothing else has or could. Nazism was a specific attack on the values that the academic community held in highest priority, and the reality of its threat was brought home to American academics by a stream of refugees from Europe whose names [were] a byword among American scientists—Fermi, Wigner, Bethe, Teller, Ewald, Franck, and many lesser luminaries. Thus academic intellectuals were well prepared, emotionally and intellectually, to close ranks behind the American war effort, Natural scientists left their home universities and flocked to the war laboratories set up by OSRD (Office of Scientific Research and Development). An informal and extremely effective system of recruiting for the war effort was established within the academic community.

At the same time humanists and social scientists flocked to the Office of Strategic Services and to the various agencies set up to manage the war economy.

As the war drew toward its close there were leaders, both in the legislative and in the executive arms of the government, who were concerned lest the vitality and momentum of the wartime research be lost in the postwar years. This is not surprising since there had always been a clear understanding that the activities of the Office of Scientific Research and Development would be completely terminated at the end of the war. Thus, the question of a continuing relationship between science and government was discussed at high levels of the government before the end of the war. One of those who addressed the question was the Secretary of the Navy, James V. Forrestal who, in his annual report to the President in 1945, identified the encouragement of research by the Navy as one of the problems worth reporting to the President. He said, in part:[3]

> In peace, even more than in war, scientists owe to their nation an obligation to contribute to its security by carrying on research in military fields. The problem which began to emerge during the 1944 fiscal year is how to establish channels through which scientists can discharge this obligation in peace as successfully as they have during the war ... The Navy believes the solution for this problem is the establishment by law of an independent agency devoted to long-term, basic, military research, securing its own funds from Congress and responsive to, but not dominated by, the Army and Navy ... The Navy so firmly believes in the importance of this solution to the future welfare of the country that advocacy of it will become settled Navy policy ... The Navy feels so deeply about the importance of the solution of this problem that it requests your intervention, guidance and support on this problem, which transcends the responsibility and authority of any single department.

The Secretary of the Navy was one of many who participated in the discussion of the need for federal funding of science and of the framework in which this funding should be provided. Another participant was Vannevar Bush, the wartime head of OSRD, whose book, *Science, the Endless Frontier,* published in 1945, provided the rationale and stimulated the drive for the establishment of a National Science Foundation. (It was not until 1950 that the Foundation was established.)

In 1945, like many others, I was asked what I would think of the creation, within the Navy, of an office that would give universities money to pursue basic research in mathematics. I expressed grave doubts. I thought it unlikely that mathematicians would be enthusiastic about receiving money from the government to support their peacetime research and even more unlikely that money from one of the military services would be welcome. But the plans for the establishment of the Office of Naval Research went forward in spite of my doubts, and that Office was brought into being by an act of Congress in 1946. I was invited to go to Washington to set up its mathematics program; and, after consulting with some of my wisest friends, I decided to participate in what still seemed to me a somewhat uncertain venture. Later, counterpart offices were created by the other two military services, and the three offices together carried on programs that provided wide support for mathematical research in American universities.

When I arrived in Washington in August of 1946, it was impossible to find a place to live. No apartments were available and most hotels permitted a guest to stay only 5 days. When I found one that extended its hospitality for two weeks at a time, I was enchanted. I made a virtue of necessity and, every two weeks, vacated my room and went on a trip to a leading mathematics department. On my return I registered for another two weeks.

These were the conditions under which I consulted many of the senior mathematicians of the United States. Together we evolved the first outline of the mathematics program of the Office of Naval Research. Basically our decision was to support pure and applied mathematics, statistics and computer development with its related numerical analysis to insure the sophisticated use of electronic digital computers when they became available. At least as important, we would try to establish the philosophy that the Navy would provide funds to buy time for able mathematicians to carry on their research, establishing research assistantships for the education of promising young mathematicians whose support seemed to us the key to the flowering of mathematical research in the country. Because the Office of Naval Research was not authorized to carry on an educational program, the purely educational aspects of the establishment of new fields would have to be handled by the universities.

Time and research experience for able students are, of course, components that any research project undertakes to provide, but the need for increasingly expensive equipment that characterized research in the natural sciences was not present in mathematics. In the beginning there was little else needed by the mathematicians, except, importantly, secretarial assistance. Later, Mathematical Reviews came to rely on government support; and page charges were included as essential elements of research projects. Travel support, particularly to international conferences, assumed continuing importance. But providing research assistantships for promising students, summer salaries for senior people, and released time for serious mathematicians with heavy teaching loads introduced a whole new ambience into the mathematics departments of many colleges and universities soon after the end of WWII. Some of our leading mathematicians believe even now that the introduction of this new force onto the campus was the most important influence of the new programs, for it changed the locus of power, making research people rather than administrators the determining force in setting educational goals and campus procedures. This, of course, varied from university to university. But within a few years the government's program had the effect of improving the working conditions and increasing the total annual salary of a substantial proportion of the country's research mathematicians and providing support to a substantial number of graduate students. These were, I feel sure, among the forces that made a career in mathematics increasingly attractive to able young people.

Not only did the government's programs have a broad effect on mathematicians across the nation. Some institutions, as well as some departments, changed their character under the stimulus of federal dollars. The ONR emphasis on analysis, with considerable interest in questions related to continuum mechanics, stimulated increased activity in several departments. The New York University Graduate Mathematics Department is one of the most striking in its use of the new resources, originally in support of analysis and, later, much more broadly. With the new funding it expanded its activities, assumed a new role in its university, and became one of the most distinguished departments in the country. The support of some very substantial engineering-oriented mathematical research by the Mechanics Branch which, after the first few years of ONR's life, was part of the Mathematics Division, extended the Division's influence into the engineering schools of a number of universities. One of these that exploited the new resources across the span of science and engineering was Stanford where not only engineering, but mathematics, including applied mathematics, and mathematical statistics flourished. Frederick E. Terman, Dean of the Engineering School at Stanford, who later became provost and vice president of the university, has commented, in a recent letter:[4]

> The Office of Naval Research...had a profound effect on the development of the mathematical sciences in the United States since the end of World War II. This came about as a result of the fact that in the critical half dozen years immediately after the end of World War II the Office of Naval Research was virtually the only source of funds available for the support of basic research in the mathematical sciences...mathematicians were important early participants in the era of sponsored research that began in 1946, a fact that contributed materially to the development of mathematics in the United States after World War II.

Possibly most significant for pure mathematics was the considerable expansion of the program at the Institute for Advanced Study which brought many of the most promising young mathematicians, as well as a number of established scholars, to the Institute each year as visitors. ONR support played an important role in enabling the Institute, in the years after World War II, to expand its influence on the growth of mathematics in the United States.

Support by the Office of Naval Research for research in the more abstract fields of mathematics, the type usually represented at the Institute, had been in our original planning, but authority for such support without

regard to relevance to the Navy's mission had not been made explicit
when the Mathematics Branch was established. Nevertheless, it seemed
clear to us in the mathematics branch that the argument for increasing
the number of well educated and experienced research mathematicians was
a strong one. During the war, the effectiveness of mathematicians in
handling troublesome and pressing problems had often depended not
nearly so much on the field of their research as on their quality as
researchers. The purest of mathematicians were among the most admired
and sought after in seeking answers to many urgent problems though
there were also some problems, like the malfunctioning of a rocket,
that required specific experience in a relevant field. Moreover, there was
considerable feeling among those of us responsible for the program that
our concern must be with the strengthening of mathematical research in
the United States not with fragmenting the field; and we wanted very
much not to exclude any first class research.

One night early in my tenure I was sitting at my desk, working late,
when I was joined by the military officer whom the staff of the research
division identified as the spiritual father of the Office of Naval Research,
Capt. Robert Conrad. He was a great man and a great leader, and his
energy and enthusiasm set the tone of ONR. He sat down, and said to
me, after a little chit-chat: "Mina, if you want to include pure mathematics
in your program, I'll support you in your decision." This was a great day
for all of us, for it meant an end to the constant worry as to whether
the Navy would see the needs of mathematics as we saw them.

The ONR program that developed had undoubted advantages for
mathematics, but it had its disadvantages, too. Since research assistantships unlike fellowships were associated with the research of a particular
member of the faculty, a student might be lured because of the availability of support to work in a field not really of interest to him. The
same is true now, of course. Moreover, the fact that the one who pays
the piper calls the tune was of concern then as now. Many, and perhaps
most mathematicians supported by the Navy, continued to work in the
abstract parts of mathematics that were of greatest interest to them.
But there were others who, to quote A. W. Tucker of Princeton,[5] "felt an
obligation to reach out beyond customary courses, seminars, and research,
to make two-way contact with industrial labs and government undertakings." My own evaluation is that those who were lured into new
fields by the Navy's interest were mathematicians who welcomed a reason
for exploring new aspects of work they had been interested in for a long
time. When this kind of new research commitment was accompanied by
the offering of related courses and seminars at their universities, a lively
campus activity was apt to come into being. Thus Solomon Lefschetz set
up at Princeton a broadly based program in differential analysis that
provided a home for the work of a number of vigorous young mathema-

ticians who, in their subsequent careers, became leaders of new developments in such areas as stability theory of differential equations, mathematical theory of control processes, and dynamic programming. And the project in the logistics program (about which I will say more later) under A. W. Tucker produced many of the leading figures now operating in universities, industry and business in fields related to the project. As George Dantzig observed:[6] "...Tucker's interest in game theory and linear programming began in 1948. Since that time Tucker and his former students (notably David Gale and Harold W. Kuhn) have been active in developing and systematizing the underlying mathematical theory of linear inequalities. Their main efforts, like those of a group at the RAND Corporation..., have been in the related field of game theory." One of the results of their work was a series of volumes in the *Annals of Mathematics Studies*[7] reporting contributions to the theory of games and answers to some of the questions raised explicitly or implicitly in John von Neumann's and Oskar Morgenstern's pioneering work, *Theory of Games and Economic Behavior.*[8] Moreover, because linear programming has become so widely important in applications in fields other than mathematics, it has assumed a particular interest as mathematicians look for cognate fields in which to seek careers for their students who can no longer expect to rely on teaching mathematics.

The use of government funds to woo scholars into research that has been identified as of national interest is a question on which there has been massive argument and disagreement in recent years. In particular the support of basic research at the universities by the Department of Defense has come under severe attack. But in the years immediately after World War II, basic research at the universities relied heavily on support by the military. More recently a number of civilian agencies with pressing social missions have entered the field. And it is interesting to note that the National Board on Graduate Education encourages this method of influencing research efforts at our universities. In its discussion of the problems currently raised by the very large number of Ph.D's being awarded in the United States, the Board argues that the United States should not rely on the types of quota system characteristic of the Eastern European countries to determine the number of Ph.D.'s to be awarded or the fields in which students may work, but should rely on the labor market to make this allocation. But then it has this to say:[9]

> ....there are instances where market forces will not produce the research and trained manpower in the volume and with the required characteristics in time to meet social needs. For example, the federal government may embark upon a large scale program to develop alternative energy sources, requiring new clusterings of research talent and advanced training facilities. The long lag that would occur before market forces generated the necessary centers for research and training would impose heavy (and unnecessary) costs upon society.

In the past, purposeful efforts to combine a stimulus to research with training programs have been most prominent in the biomedical field. New areas of research—such as molecular biology, biophysics and steroid chemistry—have been stimulated by federal grants, which provided in a single package funds for fellowships, specialized research equipment, research supplies and renovation of facilities, as well as salaries for research assistants, postdoctoral researchers and for faculty. Such support, distributed on a competitive basis, is a powerful and efficient means for hastening the development of a field of investigation in order to create new centers of strength. This mode of support can be adapted to other areas.

There are now urgent pressures to produce information, ideas, and experts in fields where the knowledge and the trained people pre-requisite to a solution do not exist. New knowledge and trained people are urgently needed to deal with the problems of energy supply, conservation and distribution; the full array of difficulties that afflict our cities; including special problems of urban housing and transporta-tion, the problems of racial tension and conflict, and the delivery of health care. Obviously, more than new knowledge and trained people are needed to solve these problems, but they will not be solved without them. Government action is needed, as it has been taken over the past two decades with conspicuous success. Institutions of higher education have reacted and adapted to national needs as expressed through the actions of Congress in passing federal laws and appropriating funds.

Under these circumstances, a set of federal programs in specifically designated, limited areas is required which will give support to the research and advanced training base of the fields in question.

It is true that the National Institutes of Health, in their Training Grant Program, have exploited this method of building into the university life of the country research and training in fields important to their mission. But so have other mission-oriented agencies of the country, conspicuously NASA. And civilian agencies like the Environmental Protection Agency and the Department of Transportation now have research support programs which, while they open up additional funding for academic research, are designed to produce research findings of importance to the mission of the agency as well as to train young men and women interested and able to assist in promoting the agency's purposes.

Thus thirty years of federally supported research have served to develop a pattern of operations, along the lines suggested for the Navy by Secretary Forrestal in 1945, that expands the ability of the nation to call on the contributions of scientists and scholars to the advancement of the national purpose as defined by Congress. Even the National Science Foundation has developed programs to encourage scientists to consider ways in which their research can contribute to the solution of national problems.

In the early days, operational problems that now loom large were present in embryo and I shall mention some of them. I have described my concern about the appropriateness of support for research in abstract

fields of mathematics. Part of our justification for this support was the
need for additional able young mathematicians. This need is no longer
present; and the increasing disenchantment with large military budgets
and with military sponsorship of work at universities has tended to diminish
the Department of Defense investment in what COSRIMS,[10] some years
ago, christened "core mathematics."

What sort of mechanism should be used to select the projects that were
to receive support? Peer Review, in its formal sense, appeared in the first
Mathematical Advisory Committee provided by the National Academy of
Sciences at the request of ONR to select those of the young mathematicians, usually new PH.D.'s, who were to receive "contracts" for a year
of postdoctoral research. The rest of the program was determined after
considerable consultation between investigators and ONR staff and was an
attempt to insure that mathematicians carrying on important research,
along with their colleagues and students, would receive support. Evaluations
were informal, and since the amount of money was small by today's
standards, congressional interest was correspondingly small.

There was no pressure for geographical distribution, though the ONR
mathematics program made a conscious effort to seek out good mathematicians in sections of the country that did not boast a leading mathematics
department when there seemed to be a good chance that a stronger
mathematical activity would develop at their universities with modest Navy
support.

Another important question that is central to the whole field of government support of university research by "mission-oriented" agencies was
with us in the beginning and continues to trouble many a member of the
university community, "How should a university faculty member respond
to the lure of Navy support for research in a field that is not precisely the
field in which he has chosen to work?" And, for those of us who were
responsible for funding in ONR, the most important question was, "How
can the Office of Naval Research maintain the enthusiasm of the Navy
for providing some of its precious funds for the support of research instead
of, as one admiral put it, 'taking three destroyers out of mothballs.' "

All these questions have their counterparts today, though the size of
the problem has been greatly magnified with the passage of years.

In the early days, we recognized that, until a National Science Foundation
was established, ONR had a special obligation to provide for the balanced
support and growth of mathematical research in the United States, always,
of course, within the framework of the Navy's established policy.

In time it became clear that dedication to these purposes would also
provide the Navy with access to first class mathematical talent to aid in
the attack on major problems. Thus, when the Defense Department
needed help in considering the DEW Line defense of the continent against
air attack, a number of our computer mathematicians were asked to give

advice. When the fleet needed a coordinated system of defense, similar invitations were extended to appropriate mathematicians. And when, in the late 1940's the staff of our office became aware that some mathematical results obtained by George Dantzig, who was then working for the Air Force, could be used by the Navy to reduce the burdensome costs of their logistics operations, the possibilities were pointed out to the Deputy Chief of Naval Operations for Logistics. His enthusiasm for the possibilities presented by these results was so great that he called together all those senior officers who had anything to do with logistics, as well as their civilian counterparts, to hear what we always referred to as a "presentation". The outcome of this meeting was the establishment in the Office of Naval Research of a separate Logistics Branch with a separate research program. This has proved to be a most successful activity of the Mathematics Division of ONR, both in its usefulness to the Navy, and in its impact on industry and the universities. Two recent Nobel Laureates in economics, Kenneth Arrow and Tjalling Koopmans, have contributed to the effort.

The original head of this branch, Dr. Fred Rigby, now Director of Institutional Research at Texas Tech University, wrote to me in response to a recent question:[11]

> We did support the quantitative side of economics substantially, for the sake of its concern with the decision making processes...[A]gricultural economics has been very strongly affected by the availability for application of the "decision mathematics" fostered by our program. We did indeed influence the introduction of operations research into business schools. The subdiscipline called management science is our invention, in quite a real sense. That is, we and our contract researchers recognized its potential, planned its early growth, and, as it turned out, set the dominant pattern in which it has developed. Naturally, we were not alone in this; RAND people in particular were also strong contributors. On the other hand, this is not just an interpretation after the fact. I recall conversations in my office that were quite specifically concerned with recognizing and fostering the new science. (It might have been better for management science had it not been so heavily dominated by mathematics, but I'm not at all sure that could have been prevented.) As you know, operations research lives in other parts of the university than business colleges. There are departments of operations research, of course, but this topic is also a major curriculum component of both industrial engineering and computer science. Nearly all of the operations research content of these disciplines derives from research areas that we supported, often as pioneers. Our journal, the Naval Research Logistics Quarterly, is a highly respected and extensively used reference for source materials in these fields. We supported research on game theory and such related topics as bargaining, ...Game models and the like have penetrated political science curricula quite notably in recent years. On the empirical side, gaming is a rather widely used instructional technique, mostly in business administration fields. Thus far I've been writing about the optimization theory aspect of the ONR Logistics Program,

and I may as well wrap that up by noting two opposite trends affecting mathematics in universities. One is for the mathematical aspects of optimization to find homes in the application disciplines and to be neglected, or at least little respected, in mathematics departments. The other and more recent is for mathematics departments to establish and operate subprograms specifically designed to service the needs of students in the behavioral science areas. Never mind the social and cultural whys and wherefores of these trends. Neither would have been what it was and/or is without the subject matter results of logistics research.

Research and applications in Logistics continue at a significant level even to the present.

Another aspect of the early work of the Mathematics Branch that not only made the Logistics program possible but touched in the broadest way the operations of the whole Navy as well as that of our whole society was the work in computers. When I first went to Washington in 1946, there were some projects dealing with analog computers that were being supported by the ONR but little if any attention had been given to automatic digital computers. John Curtiss had just gone to the National Bureau of Standards, and was eager to join forces, ONR contributing the money and participating in policy guidance, and the Bureau of Standards playing an active role in pressing forward work in the development of digital computers. C. B. Tompkins, who was then associated with Engineering Research Associates (which later became part of Sperry-Rand) was also full of ideas about ways in which ONR should participate in the emerging field. Though, with the passage of years, we did contribute in an important way toward the building of machines, our chief emphasis, at least in our original planning, was on the development of the mathematical results that would be needed if the machines, when they were developed, were to be used properly. Thus, the first ONR project I find reported at the Institute for Advanced Study is the von Neumann - Goldstine project on Methods for High Speed Computing; and a project at Columbia, of which F. J. Murray was principal investigator, was called Numerical Methods and Error Theory. There were many others on Numerical Analysis in special fields of application. In 1947 the Institute for Numerical Analysis was established at UCLA under the auspices of the Bureau of Standards and with ONR financing. (Later, the Air Force also participated). The Institute attracted to its staff, for longer or shorter stays, distinguished mathematicians and mathematical physicists from the United States and Europe. D. H. Lehmer, who was Director of the Institute for a time commented in an interview at the Smithsonian Institution in October, 1969:

> ...we were apt to do problems in support of the group that we had there, and we had a very fancy group of people: Lipman Bers, for instance, and the really good analysts and algebraists of that time came

through the INA and stayed for a month or a whole year in some cases.

So, the first priorities went to supporting their research and that's why a good deal of pure mathematics on experiments in methods and so on, partial differential equations and linear equations, solving systems and that sort of thing, was worked out there. I have a big file of things that were produced at the Institute during my tenure...the Institute was not really in business to supply service to government agencies.

When Professor Lehmer was asked how that happened, since the Institute was, after all, a part of the National Bureau of Standards, he replied in part:

> Oh, yes, but the Bureau didn't have much money in it...it was a new idea for the Bureau, really. Mina Rees of the ONR in those days was very influential in promoting this idea of getting mathematical research done at a reasonable price...the Bureau of Standards had no money tied up in the thing, really...it largely came from the Air Force and the Navy...and they...wanted this kind of work done.
>
> We trained a lot of very fine people that way, too, and when the axe finally fell on the Institute from Washington, the crowd of people that deserted the ship...permeated not only Southern California, but the whole United States, and they spread the word about what could be done, how to do it, and so in that way, perhaps, to fire everybody was a good way to disseminate information. We all went somewhere else and tried to do likewise. It was a little rough on the lot of us, because the laboratories we went to or the campuses we returned to were not supplied with the hardware we needed...and it took another ten years for that to show up at every university.

In fact, there was substantial mathematical work going on at the Bureau of Standards in Washington as well as in California. Washington, too, had notable visitors who came to work there, including some distinguished Europeans. Jack and Olga Taussky Todd spent a number of years on the Bureau staff and greatly influenced the program.

Though the Institute for Numerical Analysis was one of our principal ventures in the computer field, a substantial program for component and computer development did emerge over the years. The two computers to which we gave support that had the greatest influence on the subsequent developments in the field were the computer at the Institute for Advanced Study, known well to many mathematicians because of the central role in its history played by John von Neumann and Herman Goldstine, and the Whirlwind Computer at MIT which provided Jay Forrester with his first claim to fame.

As F. J. Weyl (my successor and later the Chief Scientist of ONR) stated in an internal paper on the *Role of ONR in the Establishment of the American Computer Technology (1 March 1957)*:

> The decisive aspects of stored program computer logic, high-speed parallel electrostatic and magnetic core memories, the interaction of high-

speed small memories with slower much more capacious ones in the same
computer, and many important simplifications in programming and coding were first realized on one or the other of these machines.

The Whirlwind, originally conceived as an element in an advanced flight
trainer, after many changes and adaptations to the state of the art, was developed as a general purpose computer of unusual design. Within months
after its completion it became clear that this computer was particularly
well suited to serve as the data processing component in a prototype integrated defense system, then under development at MIT. The experience
gathered on Whirlwind during the second half of the fifties, I am told by
Dr. Weyl, laid the groundwork for the development of the computerized air
traffic control systems, the automated reservations facilities, and the computer managed learning systems of our day.

Another major field of applications of mathematics that owes much to the
pioneering support of ONR is mathematical statistics. In the next few paragraphs I shall paraphrase a report about the contents and influence of that
program provided to me by Herbert Solomon,[12] now a distinguished member of the mathematical statistics faculty at Stanford who served for some
years as head of the mathematical statistics branch at ONR.

At first the principal feature of the ONR work was a basic research program in statistics and probability at those universities which, in the late
1940's either had such programs or were developing them. As in mathematics, the people who headed these university programs were acknowledged leaders in the field, or, in some instances, somewhat junior people
who have since assumed positions of leadership. A number of them, like
their colleagues in mathematics, were not too comfortable about accepting
Navy money to do their own thing. But, with the passage of time, they, too,
came to seek it eagerly; and with support from ONR, departments of mathematical statistics flourished on university campuses, and research activity
in the field prospered in the United States.

As we look back we see that many of the research results produced in the
ONR programs were of importance both to science and, in applications, to
the Department of Defense and to other federal agencies as well as to industry. Abraham Wald's work in sequential analysis and decision theory,
well launched during World War II, was carried forward under ONR sponsorship, Feller produced a notable exposition of probabilistic methods in
his two volume treatise,[13] and a number of theoretical examinations into
applied problems such as weather modification, models in medicine, design
of experiments and data analysis were carried on separately by Jerzy Neyman
and S. S. Wilks. Junior colleagues and students of the investigators supported in the early post-war years are now prominent statisticians, and their
students, in turn, made possible the staffing of the 50 or more statistics
departments that were begun after 1950. Others hold posts in industry.

Mathematical statistics is still a field in which there is no over-supply of professionals.

Through the late 1950's ONR helped keep statistics a viable subject in a number of universities, when there was not much support for it elsewhere. With the advent of Sputnik in 1958, more interest was shown in scientific research in general and a number of agencies such as Justice and Transportation began programs. The NIH now has a large program in statistical training and research which supports students of biostatistics and provides funds for research.

An important part of the statistics program which continues to this day derives from steps taken just prior to and at the beginning of the Korean War. About 1949 the Joint Services Program in Quality Control (now Quality Control and Reliability) was established. The focus here was to continue work done during World War II on quality control and acceptance sampling and to initiate new efforts to meet new reliability problems arising in inspection and quality measurement. A number of Department of Defense Inspection Manuals still in use were developed from this research.

I have not tried to describe the extent of the ONR program in pure mathematics since the influence of the Navy was felt primarily in the availability of increased student and research support rather than in the fields chosen for research. In the fifties a number of very able mathematicians did participate in highly classified projects during the summer and, to a certain extent, research interests generated in these summer projects did find their way into major mathematical efforts. But, on the whole, mathematical research moved forward propelled by its own inner forces; and it was natural that the National Science Foundation should gradually take over the major part of the support of this research after the Foundation was established.

ONR and its counterparts in the other military services continue to play the role envisioned by Secretary Forrestal for a military agency supporting research. A congressional decision in 1969 strictly limited the types of research that might be supported by the military services in fiscal 1970, but the range of permissible research support was broadened for the following years because the Department of Defense found the restriction so undesirable. The present controlling legislation, included in the military authorization act for 1971, states: [14]

> **Sec. 204.** None of the funds authorized to be appropriated to the Department of Defense by this or any other Act may be used to finance any research project or study unless such project or study has, in the opinion of the Secretary of Defense, a potential relationship to a military function or operation.
>
> **Sec. 205.** It is the sense of the Congress that—
>
>   (1) an increase in Government support of basic scientific research is necessary to preserve and strengthen the sound technological base essential both to protection of the national security and the solution of unmet domestic needs: and

(2) a large share of such support should be provided hereafter through the National Science Foundation.

Thus, in the years immediately ahead, it is to be expected that the primary support for core mathematics will be from the National Science Foundation; but a variety of other sources outside the Department of Defense will be available to those mathmaticians who are interested in exploring the applications of mathematics in the large number of fields of social concern which have now or soon will have programs to support research that advances their mission.

References

1. National Science Policy of the U.S.A.: Origins, Development and Present Status, NS/SPS/10, UNESCO 1968, pp. 107–109.

2. H. Brooks, in U.S. House of Representatives, Committee on Science and Astronautics, Subcommittee on Science, Research and Development, Hearings (91st Congress, 2nd session, October, 1970), pp. 931–963.

3. Report of Secretary of the Navy James V. Forrestal to President Franklin D. Roosevelt, February 10, 1945.

4. Letter from Fred Terman to Mina Rees, November 8, 1974, with enclosure, Office of Naval Research and the Mathematical Sciences.

5. Letter from A. W. Tucker to Mina Rees, July 30, 1975.

6. G. B. Dantzig, Linear Programming and Extensions, Princeton University Press, Princeton, N. J., 1963, p. 24.

7. H. W. Kuhn and A. W. Tucker, eds., Contributions to the Theory of Games, 4 vols., Princeton University Press, Princeton, N. J. , 1950–59 (Ann. of Math. Studies, 24, 28, 39, 40).

8. John von Neumann and Oskar Morgenstern, Theory of Games and Economic Behavior, Princeton University Press, Princeton, N. J. , 1947.

9. National Board on Graduate Education, Federal Policy Alternatives toward Graduate Education, Washington, D.C., January 1974, pp. 49–50.

10. The Mathematical Sciences, A Report, Publication 1681, National Academy of Sciences, 1968, pp. 57–83.

11. Letter from Fred D. Rigby to Mina Rees, June 18, 1975.

12. Letter from Herbert Solomon to Mina Rees, November 24, 1975.

13. William Feller, An Introduction to Probability Theory and its Applications, 2 vols., Wiley, New York, 1950, 1966.

14. Public Law 91–441, Armed Forces Appropriation Authorization, 1971, Sec. 204.

THE HISTORY OF COMPUTING IN THE UNITED STATES

R. W. Hamming

1. Introduction. I began doing large scale digital computing early in 1945, and hence have lived through most of what is called the Computer Revolution. You might suppose that I would be ideally suited to write about it. But the book by H. H. Goldstine, *The Computer from Pascal to von Neumann*,[1] shows me that different people can see more or less the same events in rather different ways.[2] And Goldstine says that he kept voluminous records while I have not! Indeed, it is clearly doubtful that anyone who lived through the Computer Revolution is well suited to write about it because personal experiences tend to destroy "objectivity". A talent for history, as well as training in history, are also necessary. But histories of the Computer Revolution are wanted now; we who have lived through it must do what we can until more competent historians take up the task.

Not too long ago history was mainly a description of Kings and Battles and Empires. Today we seem to prefer histories that are more culturally and intellectually oriented, that discuss more about how the common man lived and thought and hoped, and less about the Kings and their Doings. Similarly, scientific history has been almost always the history of "firsts"— who did what first, when he did it, and possibly how he did it—with scant attention to those who might have been slightly later, let alone when the widespread dissemination of the knowledge occurred.

There is a positive value to emphasizing the individual in history. Plutarch in his *Lives*[3] is clearly more interested in giving the history of the individual and his greatness, or his moral weaknesses, than he is in giving mere dates and places. His aim seems to be to inspire others to reach for greatness. The usual scientific history also emphasizes the individual, and hopefully encourages others to make the heroic efforts that are apparently necessary to advance science significantly. I am not opposed to this tendency in the history of a field *when* it is written for the experts in the field, but I presume that outsiders, including mathematicians, prefer a more culturally oriented computer history for the same reasons, whatever they may be, that makes cultural and intellectual history sought after these days.

The history of dates and individuals has the great advantage of consistency. Various historians will come up with approximately the same dates. But this consistency can be deceptive. I have attended at least two computer dedication ceremonies where the following week the computer was extensively dismantled. What, then, does the dedication date mean? At first glance it seems to mean the computer went into useful production, but clearly this can be false. The consistency and reliability of dates does not reflect what we want to know about the history of computing and of individual machines.

By contrast, the personal memory kind of history is very inconsistent, and various participants come up with rather varied descriptions of what happened. But such histories may, nevertheless, be closer to the "truth" (whatever that may be) when one is writing the history of ideas.

In spite of these difficulties, and others, I shall emphasize more than usual the general availability of computing to the scientific populace at large, as well as the growth of the central ideas of the field. I will ignore the accounting uses of machines since it falls outside your interests, and in practice it gave rise to very few significant ideas. This is not to say that the accounting use of machines was not economically important.

For most of the past 25 years approximately 70% of those installed have been IBM machines, and this means that IBM will dominate the discussion. This does not imply either approval or disapproval of IBM computers or the corporate policy; I am merely trying to measure the availability of computing to the masses of scientists and engineers. All too often a "first" was built as a single copy, and was available to a limited circle of people.

The history of ideas is very difficult, and rapidly becomes a matter of opinion. You are all, I am sure, familiar with *The History of the Calculus*[4]. The Greeks used circumscribed and inscribed polygons to estimate the area of a circle. Are we to assume they understood the upper and lower Riemann sums? Isaac Barrow, who was Newton's teacher at Cambridge, included in his lectures a proof of the fundamental theorem of the Calculus[5]. Fermat and others were aware of the method of tangents. Yet we customarily attribute the discovery of the Calculus to Newton and Leibniz. Evidently there is more to "understanding" than mere awareness of possibilities. Indeed, most mathematicians have had the experience of knowing a theorem for many years and then suddenly realizing what the theorem is all about!

The fact that the cultural history, especially of ideas, is far harder to write than is the conventional history of "firsts", and is necessarily less precise in its statements, does not mean that we should not try to meet what is clearly desired by outsiders of the field. It is usually better to approximately solve the right problem than it is to solve the wrong problem elegantly and exactly.

How, then, can I proceed? Some searching of relevant documents is essential to correct errors and supply definite dates. But the topic of the availability of computing is, as I have said, a matter of judgment. My personal experience in managing both large analog and digital installations has convinced me that whether the machine is widely available or limited to a closed circle of friends depends very much on how hard the person in charge meets the needs of outsiders on their own grounds, how well he removes the "mystique of the machine", how well he "markets" the machine and its capabilities. Again, my experience is that outsiders are more likely to use the machine on important problems than is the intimate, closed circle of friends. I am reduced, therefore, to my impressions of the people I saw around the machines at various places, the people I have met over the years and which machines they used, and the papers in the literature that I saw which referred to machines. Clearly this is an impressionistic and personal history I am giving. Let me cease the apologies now, and get to the main topic, which I will divide up into hardware (the actual machines), software (the programs that are general purpose and make the machine easier to use), and the uses of computers in the general area of mathematics.

2. Hardware. Since I am giving the history of computing in the United States, I must skip over much of what happened elsewhere and begin with a difference engine built by George Scheutz (inspired by the difference engine that Babbage did not complete). It was sold in 1856 for $5000 to a wealthy American who donated it to the Dudley Observatory in Albany, New York, where I am told it was in constant use for many years. It had four differences and 14 decimal place accuracy—no small computing capacity!

In 1876, George Bernard Grant exhibited at the Philadelphia Centennial another difference engine of which he ultimately sold 125 copies. Again a lot of computing capacity available to many people.

In 1886, William S. Burroughs started the American Arithmometer Company which developed adding and listing machines.

In 1887, Dorr E. Felt developed the famous Comptometer which was widely used for additions and subtractions, and by suitable use could, of course, be used to multiply. Division must have been a bit of an arcane art to most of the users, but is, of course, possible.

Herman Hollerith noticed that the 1880 Census of the United States took $7\frac{1}{2}$ years to process, and concluded that the 1890 Census would not be completed before the 1900 Census was taken *unless* something significant was done to speed up things. He was thus led to the development of electromechanical computing machines. By 1887 he demonstrated in field trials the efficiency and practicability of his machines. And they did handle the 1890 Census as planned. In 1896 he formed his own Tabulating Machine Corporation, which in 1911 became CTR, and finally in 1924 became the IBM Corporation.

In 1911, James Powers, who worked for some time with Hollerith, formed the Powers Accounting Machine Company which by 1955 became part of Sperry Rand.

The history of the harmonic analyzer is long. A. A. Michelson developed a very large one at the University of Chicago. When he found that the partial sums of the Fourier series had a peculiar overshoot he set the stage for the explanation[6] of the Gibbs phenomenon (1899) of Fourier series. (It was known earlier, but apparently was lost in the literature).

In 1930, Vannevar Bush at MIT developed the first practical differential analyzer. It was a flat bed of shafts and gears interconnected to correspond to the terms in the differential equations being solved. The crucial step was the power amplification device he used, a string running around a spinning shaft. Pulling gently on one end tightened the string around the shaft and this supplied the power to do the work. Thus power was supplied around the closed loops of operations that the machine was doing. It was, clearly, a mechanical machine, but it was very valuable in getting specific solutions to specific differential equations.

These are but a few examples of early pioneers in computing. Evidently Americans were very active in the computer field. They were showing the famous "Yankee ingenuity", tinkering here, adapting there, making a constant stream of improvements on what was available. Without going into details let me observe that the desk calculator gradually developed from an automatic add and substract with hand control for multiplication and division, to a fully electric powered, four function machine which was widely available by 1937. Some even incorporated square root. These were all essentially mechanical machines, though they used electric motors to drive them. And all were, basically, decimal machines.

In 1935, IBM began manufacturing the IBM 601 multiplying punch which had an effective speed of about one operation per second. In total they manufactured over 1500 of these machines, and since they were rented and not sold you can be sure that they were in fairly regular use. The 601 was essentially a relay machine, but in places it used rotating cams to close contacts.

George Stibitz in 1940 exhibited at the Dartmouth, New Hampshire meeting of the American Mathematical Society a remote console connected to the Bell Laboratories *Complex Number Computer* in New York City. It was a relay machine that did arithmetic on complex numbers. Stibitz started the development of increasingly powerful relay computers until by the Model 5 it was a general purpose computer. The Model 6 had a very extensive subroutine system.

In 1939, Howard Aiken of Harvard, in cooperation with IBM, began what was in 1944 to be the Mark 1 computer. It was a combined relay and rotating cam machine, with fairly fast operations for an electromechanical machine. He went on to develop a number of improved machines.

The appearance in 1944 of the famous ENIAC with its 18,000 vacuum tubes and all electronic insides is often taken as the start of the Computer Revolution. Certainly it could compute much faster (about 5000 additions per second) than any electromechanical machine. But when one considers the number of electromechanical machines in operation at that time, it was not a great increase in the computing capacity of the country.

The ENIAC was conceived, designed, and built by Mauchly and Eckert at the Moore School of the University of Pennsylvania with the aid of Government money, in this case the Ballistics Research Laboratory at Aberdeen, Md. where trajectory calculations were needed in large volume. In the summer of 1946 Mauchly and Eckert held a summer conference to tell others what they had done and what they thought could be done. As is so often true, they had, even before the completion of one machine, the idea of a significantly better one, the EDVAC. Many people got their first real awareness of what computing machines were all about from this summer conference. The birth of modern computing dates from around this time.

Von Neumann wrote a report under contract with them which is apparently the first public description of internal programming of a computer—there is little use in building a fast machine that must constantly wait on slow humans for further instructions, or which like the early ENIAC was rather rigidly bound to a fixed routine of computation with little chance to alter what is being done. The father of the idea of internal programming is open to debate. Von Neumann went on to propose a general purpose computer for the Institute for Advanced Study, called the IAS machine. Their efforts and reports inspired many others to build machines, and it is a curious fact that many of the copies were built and running long before the parent was.

At this point in history,[7,8] it was mainly individual Universities, often supported by Government money, who built one or two of a kind machines, and kept the field in a ferment of activity.

The Mauchly-Eckert effort finally produced the UNIVAC, the first commercial computer and installed it at the Census Bureau in 1951.

Meanwhile IBM had been slowly developing its old 601 multiplying punch into the 602, the 602A, the 603, and finally the 604 which was first delivered in 1948. The 604 was a general purpose electronic calculator in a restricted sense, being essentially card programmed with an extensive plug board wiring to suit the particular problem. More than 4000 were built.

The IBM Card Programmed Calculator (the CPC) was inspired by demands from Northrup Aviation that the IBM multiplying punch and tabulator be interconnected. When others heard about this they raised such a hue and cry that IBM in 1948-1949 met the demand by combining parts of four machines into one CPC. At one time there were some 200 CPC's in operation.

During the war Watson, Sr. of IBM had promised to build a Defense Calculator, which was to be all electronic, but after the war this was changed into the IBM 701 computer. There were 18, 19, or 20 such machines built, depending on which source you use. Although the UNIVAC was first in the field of commercial use, and had about 7 installed when the first 701s became available, IBM's superior selling ability soon eclipsed the UNIVAC and other competitors. IBM next developed a magnetic drum 650 computer which was the real work horse of industry during the 1950's, there being over 1000 installed.

IBM continued to dominate the market of computers, developing the 704 (January 1956), the 709 (1958), and the 7090 (1959) which was a transistorized version of the 709. Others, of course were making similar, sometimes better, sometimes worse, machines but none managed to challenge IBM in machines installed, and hence in computing power in the field. IBM also, beginning with a series of seminars in the 1940's, gradually dominated the field of applications.

I have not mentioned the fastest machines of their time, the IBM NORC and Stretch, the REM-RAND LARC, the Control Data 6600 and 7600, and the Cray I machine, let alone the sea of Maniacs, Whirlwind, Seac, Johniac, etc. These machines were often pioneers in some phase of construction of hardware, but in the history of use, since they were either a single machine, or at most a few copies, they played minor roles.

What were some of the ideas behind the changing computers? The early IAS design essentially had every number pass through the accumulator of the machine. Index registers, which are simple adders, were soon added to machines (after first simulating them in software) and they took a large burden off the accumulator bottle neck. Further parallelism of processing made computers significantly faster. The idea of indirect addressing, the address given *contains* the address of where to find the next item (number or instruction), is another idea that has proven useful.

There has been a gradual trend towards the situation where the computer is completely self-aware; a computer can now usually find out what it is doing, and what its current state is. This was not true in the early days. Another trend has been away from direct reference to specific parts of the machine; the machine now can assign its own choice of tape units, storage places, etc. The name "virtual storage" typifies this trend away from the specific towards the general. And we can expect this to go on in the future.

3. Software. There is a tendency for historians to concentrate on the hardware of computers. In my opinion this is the same mistake that is made by those who believe a book is its physical realization (hardware) rather than the ideas expressed therein. We all agree on the importance of the discoveries of paper making and printing; they are vital to the wide dissemination of learning—without them where would we be?—but I have not noticed that the contents of the earlier hand written books were that much worse than the average book produced today. The importance of these inventions is that they greatly reduced the cost of books, and the importance of the computer in the long run is that it greatly reduced the costs of symbol manipulation. The computer itself is a very interesting technological device requiring a great deal of careful engineering—but so is book production. I do not feel that to an audience of mathematicians I need to belabor this point as much as when writing for the average audience.

While not all the main ideas in the field of software originated in America, we have dominated the field. The reasons are not hard to find. As one European remarked to me, "When you have a large high speed storage device available, it is not hard to think about the corresponding software." That, and the fact that we have had so many more machines, are probably the main reasons for our leadership, on the average, in this area.

Computer Science began when it was clearly recognized that computers are symbol manipulating machines, not just number crunchers. But like any concept of broad general application it is difficult to pinpoint the discovery and general understanding of this idea since there were many precursors. As earlier remarked, the history of the Calculus is full of early, incomplete starts.

Similarly, in Physics we do not feel that Max Planck at first understood the quantum mechanics, nor did Einstein when he discovered the photoelectric effect. While opinions differ, there is some consensus among physicists that "understanding" occurred when Max Born observed that it is the square of the absolute value of the wave (or state) function that assigns the probabilities which can be identified with measurements. Thus the whole formal structure of quantum mechanics that had been erected was made relatively clear.

Returning to computing, when did the concept of the computer as a symbol manipulating device become clear? Did Turing in his famous paper of 1936 understand this clearly? Did Post? Do the famous reports of Burks, Goldstine, and von Neumann[9] reveal that they understood?

For myself I learned it slowly and painfully from Appendix D of the book by Wilkes, Wheeler and Gill[10], published in 1951. Did they understand clearly what they buried in Appendix D, where they showed how an interpreter could be built for a computer to translate from an almost arbitrary language into machine language? I doubt it. In each case I feel they are like the anticipators of the Calculus, they understood partly, but not clearly as we now do. By 1954 it was a clear concept to many people— I was not alone in consciously designing and building a higher level language to be translated into machine language (for the IBM 650). I had earlier done much the same, but *without* the same clear understanding, on the CPC, by wiring plug board panels.

I suggest that this central idea, that the computer is an abstract symbol manipulator, not just a number processor, became clear to many of us during the years 1952–1954. I know of no one who can claim exclusive credit— it was in the air as was the Calculus in its time— and we mutually taught each other.

To explain a bit of the jargon of the field, an interpreter translates and executes the words of the source language into machine language each time it comes to them; a compiler translates once, and then the translation is later used to drive the computer. The first is clearly inefficient of machine time, but the second leaves many problems for the user when the errors, bugs, etc., arise and he is in the translated language. From our point of view both translators and interpreters are much the same— they both allow the user to write his program in a suitable language and to leave the details of the translation to the machine. Of course the early machines were used to translate from binary to decimal and back, but as in the Calculus case, I am reserving the dating of the idea until it was quite clear that people *by their actions* understood that the computer was essentially a symbol manipulating machine.

Once we had this fundamental understanding of what the computer was, the rest of the development followed easily. What is today often called "machine language" is in fact far removed from it. It is typically a one-word-to-one-word translation much of the time, but it is far from the absolute binary programming we first labored through. Without it we would be bound to the machine like slaves; with the concept we are freed from a sea of details of programming.

Returning to the Calculus analogy, once Newton and Leibniz established the concepts clearly, it was inevitable that the vast plain now called "Advanced Calculus" would be explored. Similarly, once we grasped the idea of languages and the symbol manipulating aspect of computers, it was inevitable that the monitors, various programming languages, etc., that constitute the modern software of the computer would be developed. The details were not certain, but the general direction was clearly set.

In the Calculus thoughtful people in time began to wonder about the soundness of the practice in the field, and to create a body of theory to explain the formal manipulations of the Calculus. In computing we are beginning to create theories to explain and elucidate many of the formal, practical things we have been doing. And we can hope that in time there will be a similar benefit coming back from the theory to the practice so that we may indeed understand what we are now so frantically doing just to keep up with the pressure for the initial exploration of the domain. In my opinion we are making definite progress along these lines.

The cost of initially developing the software for a machine often greatly exceeds the design cost of the machine itself. But the duplication of software is much easier than the duplication of a machine! Software costs are still too high, but are not unreasonably expensive considering their value. We expect to reduce their costs through the application of sound engineering practices adapted to *ideas* as contrasted with *things*. The growth of software for a machine has been from a few thousand machine words to over a million, and is headed even higher. The software is what makes computers relatively easy to use. As a friend of mine once remarked, if you want to go fishing and are offered a rowboat or a battleship, take the rowboat unless the battleship is fully manned. Thus most of us judge a computer first by its software, and if that looks good we then look at the machine behind the software. We have been burned too often doing it the other way.

4. The Use of Computers in Mathematics. We now come to the use of computers in mathematics. I shall ignore other applications since they are probably generally familiar to you and are of less interest.

Early in the history of computing the great von Neumann preached that the many numerical solutions of particular cases we could compute would shed a great deal of light on many parts of mathematics, and would prove to be a significant stimulus to the *whole* of mathematics. As a very minor disciple I confess that I preached much the same message. I had known that L. E. Dickson had for years kept one or two desk calculators running to check old conjectures for counterexamples and to provide tables for new conjectures in Number Theory. I was also well aware that practical mathematics, with its emphasis on definite numbers, had in the past greatly stimulated many branches of mathematics.

But we found that most mathematicians simply ignored computers, if they did not deliberately flee from them! From the present position I think that they were more right than I was—computers have had a great deal less influence than we had hoped. Even the fundamental idea of an algorithm, which is so central an idea in computing, has not penetrated very far into the general mathematics curriculum (from computing; logic has had some effect).

Let me be specific. Early in the history of modern computing we tried proving theorems with computers—a branch of Computer Science called Artificial Intelligence. There were some astounding early successes. The famous routine for proving theorems in high school geometry produced the proof that given an isosceles triangle the base angles are equal. It used the elegant proof of comparing the triangle to itself flopped over, and concluded that corresponding angles are equal. The proof was not known to the men who wrote the program, but was in the mathematical literature. Most mathematicians agree that the proof is elegant. We even had races to see how fast a machine could prove the first few hundred theorems of Principia Mathematica of Whitehead and Russell. We started gaily down the path of a "general theorem prover" which supposedly would require only the details of a particular field to prove theorems in that field. We have yet to prove a significant new theorem using the Artificial Intelligence approach.

We have had minor successes to be sure, but our failures are too great to ignore. Work in this area is carried on by a few brave souls, but compared to the great enthusiasm we once had it is marginal. Chess playing routines are getting better, but apparently because both the game of chess and programming are better understood, rather than from great progress made after the initial wave of discoveries in the field of general game playing.

Number Theory was one of the earliest fields of mathematics to be attacked via the new big computers. A lot of machine time has been used on Number Theory problems, but the results are, in my opinion, not as much as we hoped, though certainly better than in theorem proving. Perhaps it was because their goals were more modest that their successes were the greater.

To get down to details I will make a definition that a theorem must involve an infinite number of cases[11]. This definition is, of course, arbitrary, but it has the merit that it agrees with common experience. For example, the statement that the two thousandth digit of pi is a certain digit is not a theorem, nor are statements about isolated Mersenne numbers. No patient enumeration of a finite number of cases, regardless of the amount of computing, is by this definition to be called a theorem. There must be a potentially infinite number of cases.

In proving theorems in Number Theory we must bridge the gap between the finite capacity of the machine and the infinite demand of the theorem. One technique, with many variants, is to show that as n gets larger and larger, runs of integers with a certain property get longer and longer and that ultimately the runs completely overlap and leave no gaps. Finding where this occurs, and examining all the cases up to that point is the work of the machine. It is the cleverness of the mathematician that proves the theorem. Techniques like this, and others, have been used a number of times, but each problem requires careful thought before approaching the machine. It is not as yet a mechanical process.

Have the results found in Number Theory from all this computing greatly changed the field or produced any great new insights beyond what one would expect from a giant computer? It is hard to get a clear reading from those who are deeply involved in the field. My outside opinion, after asking a lot of pointed questions and discounting enthusiasm, leaves me with the feeling that computers have added a lot of details, opened a few minor fields, but have as yet made no major contributions to Number Theory. Computers have been merely super desk calculators.

Another general area of mathematics that seems appropriate for computers is Group Theory and related topics. Again, with some differences, the machine has helped find many new, minor results, proved some results, and disproved some conjectures, stimulated much thought and research into new questions, but has not done as much as we had hoped.

We have also invaded the area of formal mathematics with programs like ALTRAN which do formal symbol manipulation as directed by the user.

The software assigns the storage, keeps track of where things are, does the brutal labor of manipulations, keeps the coefficients in rational form to avoid roundoffs, etc. And this is no small thing when you face a sea of manipulation. But again, originality seems to elude us, we have found pitifully little that is genuinely new. We can, of course, check algebraic conjectures in more cases than by hand. But in practice the storage demands, and time needed, rapidly approach infinity. This has caused some research on how to carry out certain algorithms without having things grow astronomically, but the results were mainly human generated, not machine produced. The machine does not find how to economize in significant ways.

I have confessed to you that in four of the many fields of mathematics where we have tried to use computers to do mathematics I am disappointed. You can mark it down partly to unjustified hopes that we had in the early days. According to Turing machine theory a simple universal machine can do anything any machine can do. We thought that because the theorem is true we could do all things and that we were some kind of demi-gods. Our failures suggest that we should not ignore the following remark:

Just because a human cannot program a computer to think,
does not mean that a computer cannot think.

It is time to consider that our failures to use the machine to create significant mathematics may not be the fault of the machine, it may be our own limitations. Perhaps there are thoughts that we cannot think! Certainly, there are sounds we cannot hear, odors we cannot smell, flavors we cannot taste—why not the possibility that, being what we are, there are thoughts we cannot think? This is the kind of conjecture that one is driven to when the direct proof of a theorem eludes us—we then conjecture that the theorem is false and try the other side of the argument for awile.

Of all the fields of mathematics where computers have been used probably they have done most for differential equations, especially partial differential equations, where from computed solutions people have been led to new insights. A little of the von Neumann belief has been at last realized.

Well, what can you expect from computers in the near future? Cheaper, faster, larger, more widely available computing power, more awareness of the central role of algorithms both in Computer Science and in Mathematics. But I doubt that in the next 25 years we will be doing significant mathematics in a routine fashion on computers. Your jobs, in that sense, are safe!

Bibliography

1. H. H. Goldstine, The Computer from Pascal to von Neumann, Princeton University Press, 1972.

2. H. D. Huskey, On the history of computing, Science, vol. 180, (May 11, 1973) 588–9.

3. Plutarch, Lives, Modern Library ed.

4. C. B. Boyer, The History of the Calculus, Dover, New York, 1949.

5. H. O. Midonick, The Treasury of Mathematics, Philosophical Library, 1965.

6. J. W. Gibbs, Nature, April 27, 1899.

7. Serrell, *et al.*, Electronic computers, Proc. IRE (May 1962) 1039–1058.

8. Saul Rosen, A Quarter Century View, A. C. M.

9. Burks, Goldstine, and von Neumann, Preliminary discussion of the logical design of an electronic computing instrument, Inst. Adv. Study, Princeton, N. J., 1946–1947.

10. Wilkes, Wheeler, and Gill, The Preparation of Programs for an Electronic Digital Computer, Addison-Wesley, Reading, 1951.

11. D. H. Lehmer, "Some High Speed Logic," Experimental Arithmetic, High Speed Computing and Mathematics, Amer. Math. Soc., 1963.

THE BOMB, SPUTNIK, COMPUTERS, AND EUROPEAN MATHEMATICIANS

Peter D. Lax

In the letter inviting me to participate in this program, I was asked to give a one hour talk "on the general subject of the development of mathematics in the U.S. during the period of World War II through the rosy days of the sixties. A more specific topic, the contribution of European refugee mathematicians in America, 1940–1960". I interpret this as an assignment to speak on the remarkable changes that took place in mathematics in America during these two eventful decades. The items listed in the title are, in my opinion, the principal tangible causes of these changes. Of course the changes would not have occurred without generous and farsighted federal financing, described in detail in another talk of this program.

It is most appropriate in this bicentennial year to express, on behalf of all refugees, heartfelt thanks to the American people for their great generosity in giving us, during hard times, homes, jobs, opportunity for doing work, and most of all for regarding us not as foreigners but as genuine, although somewhat peculiar, Americans. More thanks than words can express are due to Veblen for his untiring help to so many.

To see what changes took place we have to compare American mathematics before the arrival of refugees from Nazism with the situation after the war. Here is a list, necessarily partial, of American mathematicians who were active during the twenties and thirties; it includes some refugees from an earlier tyranny. It is convenient to arrange them according to their fields of interest, except for G. D. Birkhoff and Norbert Wiener, who were universal mathematicians and defy a narrow classification:

Analysis: Evans, Douglas, Hille, Levinson, Morrey, Morse, Ritt, Stone, Tamarkin.
Algebra: Albert, G. Birkhoff, Dickson, McLane.
Applied Mathematics: Fry, Stoker.

Geometry: Eisenhardt, Veblen.
Logic: Church, Kleene, Post, Rosser.
Number theory: Blichfeldt.
Probability: Doob, Uspensky.
Topology: Alexander, Lefschetz, Moore, Steenrod, Whitney.

The quality of the list is extremely impressive, the quantity a little small for a country of the size of the United States. Indeed, there were only three places with the requisite critical mass to sustain a really intense mathematical life—the Harvard-MIT complex in Cambridge, the University and the Institute for Advanced Study combination at Princeton, and the University of Chicago. This is not to say that all else was desert; Tamarkin at Brown exerted an immense influence; R. L. Moore had his unique school in Texas; the universities of Wisconsin and Michigan had much strength; Berkeley, Stanford and the California Institute of Technology had outstanding mathematicians on their staff, as well as first rate programs in physics and other sciences. These universities, as well as Rice in Houston, had a tradition of inviting leading European scientists for longer or shorter visits. Thus Boltzmann lectured in Stanford as early as 1905; he has written an amusing account of his experiences under the title: "Visit of a German scholar to Eldorado", see [1]. Nevertheless, the overall picture was a far cry from what we are used to today; jobs were very scarce, and many talented young people had no opportunity to reach their full potential.

The war changed all this; the role of science and technology in winning that struggle was crucial. Although the leaders of the scientific efforts were physicists, chemists and engineers, mathematicians made very substantial—and in a few cases essential—contributions in such fields as water waves, aerodynamics, gasdynamics, electromagnetic wave propagation, neutron transport, operations research, code breaking and others. The locus of the new work was partly at Government laboratories, such as the Ballistics laboratory at Aberdeen, the various Naval laboratories, Los Alamos, etc., and partly at universities such as MIT, Berkeley, Brown, New York University, etc. The speaker spent the waning years of the war at the atomic bomb laboratory at Los Alamos. The great secrecy surrounding the project, the remoteness and rugged beauty of the site in the mountains of northern New Mexico, the boldness of the concept of using as explosive an element that does not exist in nature, and the repeated, miraculously successful leaps into the dark created a heady atmosphere. We were very much aware that the bomb made a frontal invasion of Japan unnecessary, thus saving millions of lives on both sides. Everybody was painfully conscious that we were in a race against the Germans, and people noted with grim satisfaction that the great contributions of recent refugees from Hitler's tyranny, such as Bethe, Fermi, v. Neumann, Peierls, Szilard,

Teller, Ulam, Wigner, to name a few, tipped the scale in favor of the free world. If ever there was divine retribution, this was it.

There is a story about a dean at a university, the director of a substantial science project during the war, who liked to tell his staff that he was doing them a favor by not raising their salaries too high, since after the war they would have to be content with their prewar salaries, and that would be painful if in the meanwhile they would have gotten used to a higher standard of living. The dean's prediction turned out to be wrong; money kept flowing, and in almost every other respect scientific life after the war was different from the prewar era. This change came about because everyone concerned with science and science policy realised how important basic science is for technology, including military technology, and how important mathematics is for science. The wartime experience moved mathematicians several degrees closer to science and opened up a new range of problems to them. Also, scientists and mathematicians were elated by their contributions to victory over the greatest evil in modern time: nazism. They were pleased by the outpouring of support for their research, and flattered that they were taken seriously. The government started pumping money into research and education, supporting a very broad range of activities, not necessarily tied narrowly to technology. The result was a quickening of mathematical life and the rise of new centers. The immediate availability of mathematical muscle possessed by the refugees made such an overnight growth possible. A few case histories are very instructive:

There were two large mathematical war efforts sponsored by the Office of Scientific Research and Development at universities. One of them was at Brown under the direction of Dean Richardson and with the participation of such illustrious mathematicians as Bers, Feller, Hurewicz, Loewner, Pólya, Szász, and Tamarkin and many others. The other was located at New York University under the direction of Courant and with the participation of Friedman, Friedrichs, Lewy, Schiffman, Spencer, Stoker and others. The two groups had different fates after the war. The group at Brown dispersed; the reason for this most likely was the death of Tamarkin and the retirement of Dean Richardson. The project at NYU prospered, due to Courant's rare combination of talents, the availability of federal support from ONR, the presence in New York City of a large reservoir of mathematical talent ready to be tapped, and the help and encouragement Courant received from George Roosevelt, Chairman of the Board of Trustees of New York University who took a great interest in mathematics.

Stanford was another university that took advantage of the postwar boom to build up rapidly mathematics and statistics. To be sure, there was a tradition of excellence; Blichfeldt, a very distinguished and unjustly neglected figure, and Uspensky, member of the Russian Academy of Sciences before his emigration, represented a very high standard. But it was under the lead-

ership of Szegö that Stanford established itself as one of the foremost schools
of analysis and statistics. Timely support by ONR was crucial, and so was
the backing of the graduate dean, Bowker, and the provost, Terman. In
fact, under their regime during the following decade Stanford rose to great
heights in the physical, biological and engineering sciences.

At about the same time the University of Chicago reestablished its position
in mathematics under the remarkable chairmanship of Marshall Stone.

At the three universities mentioned above, as well as others that began
their buildup, refugees—alongside native Americans—filled key positions.
It is time to give a list, necessarily partial, of these illustrious immigrants;
it is arranged by fields, except for Hermann Weyl and John v. Neumann,
who cannot be classified so narrowly. I have included in this list postwar as
well as prewar immigrants:*

Analysis: Ahlfors, Aronszajn, Bers, Beurling, Bochner, Calderón,
Courant, Friedrichs, Hopf, John, Kakutani, Lewy, Loewner, Moser, Pólya,
Rademacher, Schiffer, Schoenberg, Szegö, Weinstein, Wintner, Zygmund.

Algebra: Artin, Borel, Brauer, Chevalley, Chow, Harish-Chandra,
Iwasawa, Magnus, Taussky-Todd, Zarisky.

Applied Mathematics: v. Kármán, C. C. Lin, Luneburg, v. Mises,
Prager, Synge.

Geometry: Busemann, Chern.

Logic: Gödel, Tarski.

Number theory: Erdös, Selberg, Siegel.

Probability: Feller, Kac.

Statistics: Gumbel, Neyman, Wald.

Topology: Dehn, Eilenberg, Hurewicz, Menger, Ulam.

Before describing the scientific contributions of this distinguished cast of
characters, a few words about their personalities. Certainly they were in-
dividualists, eccentricity being more tolerated on the Continent and England
than in America at that time. (Of course Norbert Wiener was as eccentric
as they come, but I am not sure how well he was tolerated.) How did this
eccentric crew fare in the land of the puritans? Most of them were eager to
assimilate. An extremely distinguished mathematician of German origin
who settled in Princeton went so far that he decided to master baseball, and
set aside Saturday afternoons for that purpose. A visitor, calling during the
sacred hour, was warned at the door by the great man's wife: "Ssh! Hermann
is listening to the ballgame."

Other immigrants tried to cling to the ways of the old world. One of them
was heard† explaining to a friend, also from Europe:

---

*A complete list of mathematicians persecuted by the nazis is given in [5].
†Attested by Sylvia Aronszajn.

"Now that we are in America we must behave as Americans, so in public you must call me Stefan and I will call you Hilda; but of course when we are among ourselves you will continue calling me Herr Professor and I will call you Frau Professor Doktor."

Such an attitude was exceptional and temporary; eventually all refugees became as American as Apfelstrudel.

What about the scientific personalities of the immigrants? They brought traditions that were very different from the prevalent ones, in style and content, thus increasing tremendously the breadth—as well the depth—of mathematical life here. In particular their greater closeness to applications, especially in physics, fitted the postwar mood very well. Many of the newcomers were in their prime and put forth their ideas with a great deal of vigor and confidence. Apparently, the tradition of showmanship in lectures was stronger in Europe than on these shores so the newcomers readily found students and disciples. It is a pity that there is no filmed record of some of the outstanding lecturers, of which perhaps Artin was the most brilliant.*

V. Neumann was a key figure in the transition from pre-war to post-war mathematics. The shifting of his own interest from pure to applied mathematics is very evident from a perusal of his collected works, [7]. Much, but by no means all, of his postwar work had to do with computers; he made many highly technical contributions to their hardware, software and utilization in scientific computing, and he speculated on their future role. For instance, in a talk in Montreal in 1946 he suggested that studying typical numerically computed solutions of nonlinear differential equations would lead to theoretical understanding. This has happened in the last ten years in a surprising number of instances, such as Kruskal and Zabusky's discovery of solitons, Ford's calculations of the dynamics of the Toda lattice which led to the proof of the complete integrability of this system.

This is no place to review the life work of v. Neumann; I refer instead to the special issue [2] of the Bulletin of the AMS devoted to that subject. Further information on the personality of v. Neumann is contained in Ulam's charming book [8], and in a very fine film [6] made under the auspices of the Mathematical Association of America. However, no mere list of v. Neumann's achievements gives a proper picture of the man; for those who are too young to have glimpsed him I offer the image of Gelfand and Michael Atiyah rolled into one, with a couple of physicists and economists added for good measure. He carried thinking farther than most people can conceive of its being carried. That is the reason he was so much sought after by the government for advice. In an interview a few years ago on Hungarian television, Eugene Wigner was asked if it were true that the U.S. Government reached many scientific decisions by simply asking v. Neumann for his opinion. Wigner in his characteristic precise manner said, "Well, that is

---

*Not all the Europeans were great lecturers; some were truly incomprehensible.

not exactly so; but once von Neumann analyzed a problem, it was clear what was to be done."

Not only the government but many mathematicians sought v. Neumann's advice about their research. Not that v. Neumann was able to solve a difficult problem in a single interview, but he had an uncanny ability of relating it to other problems. Often such a reformulation represented the labor of six months of the person who posed the question.

In spite of his superhuman ability for analysis, v. Neumann was not always right; e.g., he was against floating point arithmetic and against large core memory for computers. But his influence on the whole was overwhelmingly good and his premature death robbed applied mathematics and computer science of a natural leader, a spokesman and a bridge to other sciences.

I would like to close with a couple of social comments. It is acknowledged that before the war antisemitism was fairly common in academic circles. I would like to recall here an incident which has partly sinister, partly comic overtones, concerning the appointment of Norman Levinson at MIT. Hardy was visiting Cambridge when the matter was proposed, and being an admirer of Levinson, he gave his endorsement in person to a high administrator. He was told that there was difficulty with making the appointment because there was no room for another Jew on the MIT faculty. Hardy said he didn't realize that MIT was an Institute of Theology, and threatened to expose the affair in the pages of *Nature* unless the appointment went through—and it duly did. One indirect beneficial effect of nazism was that it inoculated a whole generation against the sickness of antisemitism, so that such behavior by a university became inconceivable after the war.

No chronicle of the fifties is complete without a mention of the McCarthy era and its effect on science. Here are a few highlights:

In 1950 the Regents of the University of California imposed a loyalty oath on its faculty. As I recall the oath was innocuous in itself; but the idea that the Regents could extract it from professors was odious. Independent minded faculty members, among them a number of scientists and mathematicians, quit rather than sign. They were fully vindicated when the courts declared the oath illegal. One can take particular satisfaction in the fact that the President of the University of California today was one of the nonsigners 25 years ago. A similar loyalty oath requirement was imposed in Oklahoma, where it resulted in Aronszajn moving from Stillwater to Lawrence, Kansas.

The campaigns of McCarthy and of the House Unamerican Activities Committee were a serious threat to the scientific community. Although there was very little sympathy for hard core members of the Communist party who have followed the party line through every twist and turn and served as apologists for Stalin's byzantine cruelties, scientists realized that academic freedom is indivisible, and that the aim of the charges was not to reveal some nonexisting conspiracy, but to make political hay for the fearless vampire hunters.

I never had to prove my loyalty in a formal security hearing, but several of my friends did. In two cases I testified for the defense; in both instances the charges were flimsy, but at least they were stated above board, and—being flimsy—were dismissed. Other victims had to fight vague, Kafkaesque charges, based on evidence in closed files. In those days the nation sadly lacked a Freedom of Information Act.

By and large, universities and especially individual mathematicians behaved honorably. We closed ranks, stood up to our tormentors, defended the unjustly accused, and scrambled to find jobs for those who were unfairly dismissed. So the long-term damage was small.

I conclude by calling attention to the interesting collection of essays [4] on the intellectual migration, and to Laura Fermi's book [3] on the same subject.

Bibliography

1. L. Boltzmann, Collected Works.
2. John von Neumann, 1903-1957, vol. 64, No. 3, Part 2, Bull. Amer. Math. Soc., May 1958.
3. E. Fermi, Illustrious Immigrants, The Intellectual Migration from Europe, 1930-41, University of Chicago Press, 1968.
4. D. Fleming and B. Bailyn, ed., The Intellectual Migration, Belknap Press, Harvard University Press, Cambridge, Mass., 1969.
5. Jahresberichte der Deutschen, Math. Vereinigung, v. 71, 1969, 167-228, v. 72, 1971, 165-189, v. 73, 1972, 153-208, v. 75, 1974, 166-208.
6. Modern Learning Aids.
7. J. v. Neumann, Collected Works, vol. I-VI, Pergamon Press, distr. by Macmillan, New York, 1963.
8. S. Ulam, Adventures of a Mathematician, Scribner's, New York, 1976.

PANEL: TWO-YEAR COLLEGE MATHEMATICS IN 1976

Moderator: R. D. Larsson

TRENDS IN TWO-YEAR COLLEGE MATHEMATICS

Donald J. Albers

1. Introduction. What is two-year college mathematics? The answer—"It is not simply the first two years of a four-year college curriculum."—comes as a surprise to many people. In the recent past, most two-year colleges were called junior colleges and most junior colleges were essentially devoted to college transfer programs. Thus the surprise mentioned above is easy to understand. Today the typical two-year college has a focus much broader than college transfer. A detailed comparison of the two-year college curriculum with the four-year college (and university) curriculum helps to illustrate the fact that two-year college mathematics is not simply the first two years of a four-year college mathematics program. In 1970–71 the Conference Board of the Mathematical Sciences (CBMS) carried out an extensive survey of undergraduate training in the mathematical sciences resulting in a publication of the same name [1]. The CBMS survey is an invaluable resource for anyone interested in two-year college mathematics and is the main reference for all of the following comparisons.

2. Remedial Courses. Remedial courses will be examined first (Table 1).

Summary for remedial courses (as of 1970-71)

33% of two-year college enrollments were in remedial courses. 7.5% of four-year college enrollments were in remedial courses. Moreover, total enrollments in the remedial area were as follows:

Two-Year Colleges:	191,000
Four-Year Colleges:	101,000

Two-year colleges carried nearly ⅔ of the *total* enrollment in remedial courses.

Trends: It is to be noted that this data is now five years old. (A follow-up

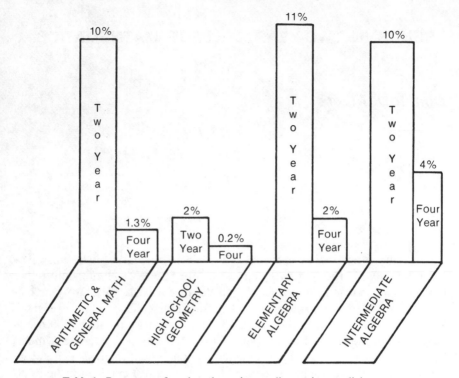

Table 1. Percentage of total mathematics enrollments in remedial courses.

to the 1970–71 CBMS survey will be released in the fall of 1976.) An accelerating decline in mathematical ability of entering college students has been observed since 1966. It is expected that the 1975–76 survey will show that remedial courses constitute a larger percentage of enrollments in both two-year colleges and four-year colleges. The fact that several campuses of the University of California are again offering courses in intermediate algebra is evidence along these lines. At a large suburban community college in California, remedial courses now constitute 50% of all math enrollments.

The Seventeenth Annual AMS Survey by John Jewett [2] also cites considerable growth in the remedial and precalculus areas at the four-year level.

3. Precalculus courses. Next consider the precalculus picture (Table 2).

Summary for precalculus courses: A standoff

21.7% of two-year college enrollments were in precalculus courses. 22% of four-year college enrollments were in precalculus courses.

4. Calculus Courses. On to calculus (Table 3).

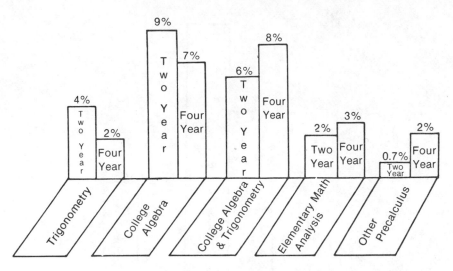

Table 2. Percentage of total mathematics enrollments in precalculus courses.

Summary for calculus courses

12.2% of two-year enrollments were in calculus courses. 26.7% of four-year enrollments were in calculus courses.

Trends: Two-year college calculus enrollments have probably increased as more "soft" calculus courses (i.e., calculus for biological, social, and management science) have been introduced in two-year colleges. R. D. Anderson in his 1975 Report on Employment Data and Academic Mathematics suggests that recent increases in four-year college calculus enrollments may be due to new or expanded courses for business students [3].

5. Elementary service courses. Elementary Service Courses are examined next (Table 4).

Summary for elementary service courses

30.2% of two-year enrollments were in elementary service courses. 22% of four-year enrollments were in elementary service courses.

Trends: In view of increased enrollments in business, we can expect to see increases in business math, finite math, and elementary statistics. (R. D. Anderson's 1975 Report indicates increases in statistics in four-year colleges.) Astonishingly, courses in slide rule are *still* given today by several two-year colleges in California. It is expected that slide rule courses will give ways to courses focusing on the use of hand-held calculators. Enrollments in courses for elementary teachers probably have declined sharply since 1970–71.

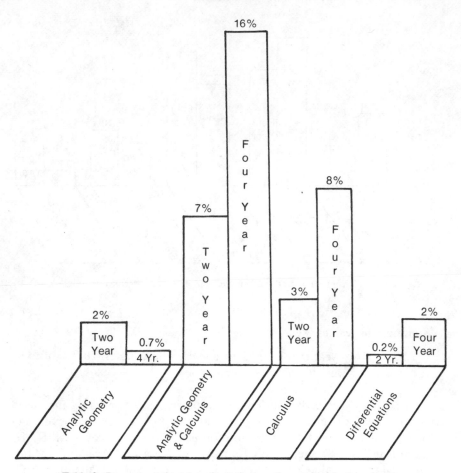

Table 3. Percentage of total mathematics enrollments in calculus courses.

Technical mathematics is an especially interesting area. From 1966–1970, enrollment in occupational and technical programs *increased* by a whopping 229% [4], while enrollment in technical math increased by only 45% [1], [7]. From 1970–74 enrollments in occupational and technical programs soared by an additional 55% [5]. Perhaps this is simply a natural lag. It appears, however, that mathematics for occupational and technical students is being done in an in-house fashion. Less than half of the 50 community colleges in Northern and Central California even offer courses in technical mathematics [6]. The dean of occupational education at a large community college in California suggests that two-year mathematics teachers need to ask occupational and technical faculty what mathematical skills are needed before attempting to institute new "tech math" courses.

Any employment projections for two-year college faculty should take into

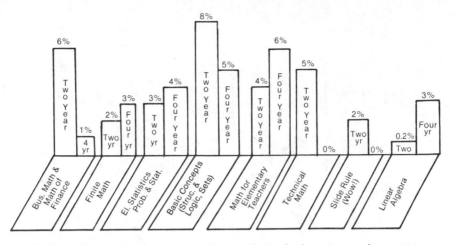

Table 4. Percentage of total mathematics enrollments in elementary service courses.

account this apparent technical math gap. The CBMS 1975–76 Survey which is now under way should be examined carefully regarding technical mathematics.

The Committee on Two-Year Colleges (CTYC) of MAA is at this time studying technical mathematics in two-year colleges. A CTYC survey carried out in the fall of 1974 indicates activity on the tech math front, particularly with regard to the production of modular-like materials.

6. Summary. We now give an overall summary (Table 5).

Overall Summary

1. Excluding courses in computer programming, 85% of math enrollments in two-year colleges were below the calculus level.
2. 52% of math enrollments in four-year colleges were below the calculus level.

Implications. On the basis of curriculum alone, one can see a great difference between two-year colleges and four-year colleges.

These differences should be made known to anyone who is considering teaching in a two-year college. Furthermore, graduate schools need to thoroughly acquaint themselves with nearby community colleges and give consideration to broadening and deepening the preparation of potential two-year college faculty. How many graduate departments require future two-year college mathematics instructors to study synthetic and metric geometry? It seems they should since 70% of large two-year colleges offered courses in high school geometry [1]. How many graduate departments require future two-year college math instructors to study statistics? They should since 60% of large and medium-sized two-year colleges offered courses in statistics [1].

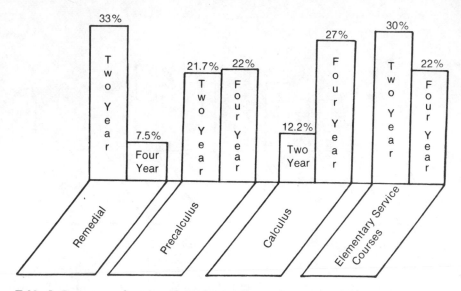

Table 5. Percentage of total mathematics enrollments in categories of courses (1970–71).

How many graduate departments require future two-year college math faculty to study elementary number theory and the history of mathematics? A knowledge of the higher arithmetic and evolution of mathematical concepts seems especially important to anyone who will teach arithmetic to two-year college students who have "seen" all of this stuff before. Shouldn't teachers of arithmetic be telling their students about some of the easily stated problems of arithmetic which are still unsolved?! How many graduate departments require that future two-year college math instructors study elementary computer science and numerical methods? How many makers of graduate requirements have actually spent a full day or two within a two-year college? One is unlikely to find the stimulation provided by a weekly colloquium nor is one likely to find someone hard at work on a problem from the *Monthly* or *TYCMJ.* Most two-year colleges do not reward research activities, nor do they support professional growth by providing financial assistance to attend meetings, to sponsor colloquia, etc.

Two-year college mathematics faculty are sometimes accused of not being interested in mathematics. If the accusation has substance, two-year college mathematics is in grave trouble, for when one's passion for mathematics is gone his teaching death is close at hand. One certainly does *not* live by teaching alone. Surely two-year college mathematics faculty initially had an interest in mathematics or they would not have pursued majors in mathematics. How does one lose interest in the Queen of the Sciences? And if one has lost interest, how can it be rekindled?

In early January the Physics Department of Stanford University sponsored a meeting of community college teachers of physics. The meeting was a smashing success. It was especially gratifying to hear the chairman of Stanford's Physics Department asking them for suggestions as to what Stanford might do to help them and vice versa. It was even more gratifying to watch the chairman *listen* thoughtfully and busily take notes on the words of two-year college physics faculty.

In addition to basic pedagogical questions, there was great interest in up-date short courses, colloquia, and the like, i.e., in being rekindled. The two-year college people were so pleased with the exploratory meeting that they asked the chairman if they could reassemble in one month to flesh out their ideas. Strong bridges were built between Stanford and surrounding community colleges on that day.

The mathematical community might consider similar one day exploratory sessions. The potential gains to two-year college math faculty and their students, and the mathematical community as a whole, seem to be enormous.

The question: "Where is two-year college mathematics going?" has already been partially answered. An important question at this point might be "Where is the two-year college faculty going?"

First consider educational attainment (Table 6).

	1966	1970	1974
Ph.D. in Math Sciences	0.4%	2%	6%
Masters—Math Sciences	56%	61%	60%
Doctorate—Math Education	1.6%	1%	4%
Masters—Math Education	21%	21%	21%
Bachelors—Math	8%	5%	3%
Non-Math Field	14%	10%	7%

Sources: 1974: [2], 1970: [1], 1966: [1].

Table 6. Educational qualifications of TYC math faculty.

Little needs to be said about Table 6 except that it is encouraging and may hold out hope for gains in professional mathematical activity at the two-year college level. The ongoing 1975–76 CBMS survey will try to make a few crude estimates of professional mathematical activity among two-year college faculty.

We close by looking briefly at an area of ever increasing concern to all of us.

8. Job prospects. The area of job projections is certainly one where even angels fear to tread. Some information that seems desirable in making projections is the following:

1. Present number of two-year college faculty (full-time and part-time)
2. Age distributions
3. Enrollment trends
   A. Overall two-year college enrollment
   B. Special areas, e.g. occupational-technical programs
   C. Mathematics enrollments
4. Number of live births over last 18 years
5. Part-time employment trends (In California, a 60% full-time, 40% part-time breakdown is emerging.)

CBMS is in the process of obtaining data on all of the above and should soon be in a good position to make some projections by fall of 1976.

To see how difficult projecting can be, consider the following data array (Table 7).

	FACULTY SIZE			ENROLLMENT		
					(In Thousands)	
	Full-Time	Part-Time	F.T.E.	Math	Total TYC	Tech. Math
1966	2677	1318	3122	348	1464	442
% Change	82%	68%	80%	68%	71%	229%
1970	4879	2213	5617	584	2500	1013
					41%	55%
1974	?	?	?	?	3527	1573

Sources: [1], [4], [5], [7], [8].

Table 7. Projections.

References

1. J. Jewett and C. R. Phelps, Undergraduate education in the mathematical sciences, 1970–71, Conference Board of the Mathematical Sciences, Washington, D.C., 1972.

2. J. Jewett, Seventeenth annual AMS survey, Notices Amer. Math. Soc., 20 (1973) 343–347.

3. R. D. Anderson, 1975 Report on employment data and academic mathematics, Notices Amer. Math. Soc., 22 (1975) 357–362.

4. M. Russo and R. M. Worthington, Trends in Vocational Education, U.S. Department of Health, Education, and Welfare, Office of Education, Washington, D.C., 1972.

5. S. M. McMillen, Vocational and Technical Education Selected Statistical Tables Fiscal Year 1974, U.S. Department of Health, Education, and Welfare, Office of Education, Washington, D.C., 1975.

6. Northern-Central Community College 1975 Mathematics Directory, California Mathematics Council-Community Colleges.

7. J. Jewett and C. Lindquist, Aspects of undergraduate training in the mathematical sciences, Conference Board of the Mathematical Sciences, Washington, D.C., 1967.

8. S. L. Drake, 1975 Community, Junior and Technical College Directory, American Association of Community and Junior Colleges, Washington, D.C., 1975.

ROLE AND STATUS OF TWO-YEAR COLLEGE FACULTY

Peter A. Lindstrom

This being our Bicentennial Year, to trace the history of the last two hundred years of the two-year college mathematics teacher in 15 minutes is not an easy task. But if one considers the middle 1890's as being the birth of the "two-year college" in the United States, that takes care of the first 120 years. The next 70 years can be summarized by saying that "nothing much happened during these years." It has only been the last decade that very much excitement has really taken place for the two-year college mathematics teacher.

Ten years ago, the role and status of the two-year college mathematics teacher was certainly not a very bright picture in mathematics education. A statement made at the MAA Board of Governors Meeting in Houston, Texas, exactly nine years ago today (January 25, 1967) gives one some idea of the situation at that time. In a resolution addressed to the National Science Foundation, mention is made that: "It appears that mathematics instruction in two-year colleges may be the weakest link in the whole range of mathematics education." At that time, this may well have been true in terms of the role and status of the mathematics teachers at the two-year college level. In many cases, these teachers simply did not have a proper training and educational background to teach at this level. In other cases, their previous experience had been at the high school or four-year college level and they did not understand the problems unique to the two-year colleges. Nearly all of these teachers had one thing in common though— very little professional identity, as few belonged to mathematics organizations and few did anything with mathematics outside of the classroom.

The past decade though has seen many positive changes for the two-year college mathematics teacher. We have seen the development of many state two-year college mathematics organizations. Many of these state organizations arose from a dissatisfaction with the MAA, the NCTM and state and local secondary school organizations. All of these new state organizations have helped to raise the level of two-year college mathematics and to improve the status of the two-year college mathematics teacher. Such state groups have given the two-year college mathematics teacher many opportunities they previously never had.

The past decade has also seen increased support from national organizations. The NCTM has aided two-year college mathematics in many ways. Many of its publications appeal to the two-year college people, portions of NCTM meetings are devoted to two-year college mathematics, the NCTM Board of Directors has a specific member from the two-year college ranks, the NCTM has two two-year college people serving as representatives to the Editorial Board of the *Two-Year College Mathematics Journal*, and one of the NCTM affiliated groups is the Florida Junior College Council of Teachers of Mathematics.

Within the past two years, we have seen the formation of AMATYC (American Mathematical Association of Two Year Colleges), a new national organization for two-year college mathematics teachers. Many people have raised the question, "With the MAA, the NCTM and numerous state two-year college groups, is there a need for another mathematics organization?" No matter how the question is answered, the future and the success of this new organization will depend upon the support of its membership.

In the past decade, the two-year college people have received much support from the MAA at the national level. Many of the CUMP publications (e.g., Qualifications for Teaching University Parallel Mathematics Courses in Two-Year Colleges, A Transfer Curriculum in Mathematics for Two-Year Colleges, A Basic Library List for Two-Year Colleges, etc.) have been directed towards the two-year college people. In 1971, we saw the first two-year college people serving on the MAA Program of Visiting Lecturers. This has grown to where there are now five two-year people involved in this activity. We have also seen two-year colleges using the services of this activity quite often in the past decade. Within the past few years, we have seen the rise of the MAA Committee on Two-Year Colleges, most of its members being from the two-year college ranks. We also see now many two-year people serving on other MAA Committees. Also, the Second Vice-President of the MAA is often a two-year college person. In the past few years, special efforts have been made to have two-year college people serve on the Program Committees for both the January and August MAA Meetings. In turn, various items on these programs are geared for people at the two-year college level. The MAA has also given support to the two-year college people with many of its recent publications (e.g., the Dolciani Series, Selected Papers on Calculus, the New Mathematical Library, etc.). Time does not allow discussion of various other items here, but let's not forget the *Two-Year College Mathematics Journal*. Founded in 1970 by

Prindle, Weber, and Schmidt, Inc., the MAA took over the publication of the journal in September, 1974, and now it is one of the official journals of the Association. Being well received by both two and four-year college people, there are now over 4,300 subscribers to the *TYCMJ*.

The two-year college people have also received support from the MAA at the section level within the past ten years. Many section officers are from the two-year college ranks, with some sections having a specific officer for the two-year college people. During this period of time we have also seen many section meetings being held at two-year college campuses. We have also seen more two-year college people making presentations at section meetings. Some sections now devote a portion of their section meetings to the problems of the two-year college people. Other sections make no distinction within their membership, preferring to integrate the two-year college teachers with its other members.

As individuals, the two-year college mathematics teachers have also made many noticeable changes in the past ten years. In many respects, they have taken on the role of being a "good teacher" and not a "publish or perish teacher." Their educational background has changed, as many have furthered their education and have obtained a background in many different areas of mathematics in order to teach the variety of courses offered at the two-year college level. We have also seen some changes in graduate school programs, the end result being people specifically trained to teach at this level. In looking at the freshmen-sophomore level math texts on the market today, we now see that many are written by two-year college people. The two-year college mathematics teacher has also taken a leading role in the development of two mathematics journals, the *Two-Year College Mathematics Journal* and the *MATYC Journal*. Both of these journals serve as a valuable means of communication for the two-year college teachers.

In short, this particular area of mathematics education has made progress in the past ten years. Even though such progress has been made, I feel that two-year college mathematics teachers still have many areas in which more progress can be made. If one looks closer at the past ten years, one sees that the progress made has been the result of the hard work of a small handful of interested and devoted teachers from both the two and four-year college levels. By no means has it been a team effort as there are presently many two-year college teachers who are contributing nothing to mathematics. These are the ones who teach their classes, and probably do a good job (thus upholding their role of being a "good teacher"), but who keep mathematics solely within the classroom. They do not identify themselves with other teachers of mathematics; they have no professional interests outside of the classroom. Not only do they hurt the image of two-year college mathematics, but this lack of interest will eventually hurt their teaching. Even though progress has been made in the past ten years,

I feel that with this apathy, two-year college mathematics is still a "weak link in the whole range of mathematics education." But is it the "weakest link?" I doubt it, as all levels of mathematics education are confronted with similar problems.

In looking at the future role and status of the two-year college mathematics teacher, I am optimistic that we can strengthen that "weak link." I look to the future as a time of unlimited opportunities for the two-year college people in terms of being not only a "good teacher," but also a mathematician. As a teacher, we must maintain our image of being a "good teacher" in the classroom, but at the same time we must not keep our mathematics solely within the classroom. We must let others know what we are doing with our teaching; at the same time, we must find out what others are doing in remedial math, technical programs, transfer programs, continuing education, etc. The future is also a time for two-year college people to realize that they can make other valuable contributions to mathematics than just through their teaching. There are many mathematics organizations at the local, state, sectional and national levels for two-year college people to join. But joining the organization is not enough; becoming an active participant and giving support of the activities of these organizations is what is needed. There are many mathematics journals that are geared to the two-year college level; more two-year college people should be subscribing to these and reading them. Also, more two-year college people should be making contributions to these journals by writing articles, reviewing articles for the editors, submitting problems and solutions, reviewing textbooks, etc. More two-year people should be attending conferences, workshops, seminars and meetings and making presentations on mathematical and/or pedagogical topics.

Professional identity and professional interests such as these, along with maintaining the role of being a "good teacher," will enable two-year college mathematics to become a "strong link," if not the "strongest link in the chain of mathematics education." The progress made in the past ten years has been the work of a few; the progress in the future must be through the work of all two-year college mathematics teachers.

THE PLACE OF TEACHER EDUCATION IN THE TWO-YEAR COLLEGE

Shelba Jean Morman

The two-year college teacher of mathematics can exert important and unique, although often somewhat subtle, influences on both elementary and secondary preservice and inservice mathematics education. I propose that where these influences can positively affect teacher education, they should be maximized. I would like to discuss these influences first from the standpoint of elementary teacher education and second from the standpoint of secondary teacher education.

In Texas it is now possible, although I would hope improbable, for one to become an elementary teacher of mathematics without ever having completed a first course in algebra or geometry. Other states show a similar irresponsibility toward elementary teacher education. Unfortunately when state departments choose to prescribe only low or minimal teacher certification requirements, this leaves colleges of these states an excuse for their own low requirements where they exist. The problem is magnified when colleges of education support and approve such low requirements. Too often, I fear, colleges also use low requirements as an attempt to compete for enrollments. That we should decry declining mathematics achievement and ignore the inadequate preparation of our elementary teachers is somehow incongruous.

The only, if any, college mathematics the elementary teacher is required to take for teacher certification is generally taken during the first two years. State requirements in mathematics for elementary teacher certification vary from zero to six hours, although there are exceptional cases of up to twelve hours. Furthermore, in some states the elementary certificate is valid for the first eight grades. These facts alone are a demand for teaching and learning efficiency wherever these courses exist. Since many of our elementary teachers will begin their college education at a two-year college, the two-year college mathematics teacher is faced with this tremendous responsibility.

Recommendations for the content of the mathematics courses for the elementary teacher come from many sources. Those of the Committee on Undergraduate Preparation in Mathematics of the Mathematical Association of America are probably the best known and have the longest history. Unfortunately, efforts to follow these recommendations are thwarted by the large number of prospective elementary teachers who are unprepared for such courses. Perhaps the competency based movement will bring about a recognition that to expect certain competencies at the completion of a course, we must look for evidence of certain competencies at the entrance. Minimally such courses should give the elementary teachers command of the conceptual tools that will allow them to portray mathematics as a living subject and not as a stagnant collection of facts. Regardless of the content of these courses it is clear that the instruction must be maximally effective and require concept mastery to some minimal level. These courses should require more than that the elementary teacher be able to rotely generate the algorithms of arithmetic. A major objective should be to give insight into the nature and structure of mathematics. Certainly a recognition of algebra as generalized arithmetic should be a minimal requirement for entrance into such courses, and minimal requirements should exist for successful exit. If such monitoring does not exist elsewhere, I am suggesting that it be done at the instruction level.

At the present, regardless of the title, the content and treatment of these courses may vary according to the whim of the instructor. Frequently, these courses are low status courses within the mathematics department and the responsibility for such is accepted condescendingly. I am suggesting that the importance of these courses, where fortunately they exist, should be acknowledged. They should be taught by specially trained persons who feel a responsibility for teacher training, who will carefully monitor the progress of each student, and who will demand high standards of performance. Evidence of the competence of the instructors of these courses must be based on more than apparent empathy indicated by a negatively skewed classroom distribution of grades. These people must be competent mathematically and knowledgeable of how the mathematics they teach relates to the mathematics curriculum of the elementary school. To send an elementary teacher out to teach addition of fractions who does not see any relationship between the addition of $\frac{2}{3}$ and $\frac{3}{4}$ and the addition of a/b and c/d, with a, b, c, and d counting numbers, is surely to be remiss in our responsibility.

The task of the elementary teacher must not be narrowly viewed. The elementary teacher needs a sophisticated level of understanding of fundamental concepts if he is to ably exploit the opportunities given him for enhancing the learning of the students as they respond to experiences in mathematics. It should be recognized that for the elementary school child the process of learning should not be mere memorization, that facts and verbalization are not satisfactory outcomes and that transfer of training is not automatic. This must be conveyed to the elementary teacher, and he must be given the knowledge and skills that will allow him to avoid the kind of teaching for which these things are the consequence.

The ever present product oriented mathematics activities which emphasize immediate learning of very specific information in a tightly controlled expository sequence are inappropriate for major emphasis in the classroom for they do not allow the needed freedom for developing one's skills in problem solving, in formulating and testing hypotheses, in learning how to learn. If we expect the elementary teacher to elicit the kinds of learning behavior from his students that we would have him elicit, he must have previously been actively involved in the same kinds of learning behavior. If we expect the elementary teacher to involve his students in activities requiring the transfer of what they know to isomorphic cases, should we not involve him in similar activities? It is probably true that teachers tend to teach as they were taught.

It behooves the two-year college teachers to be more adamant on this issue of teacher competence in mathematics. The contribution that two-year college faculties can make to teacher education may ultimately determine the success or failure of the elementary teacher in the mathematics classroom at least in so far as knowledge of mathematics determines success. The implications of this for the quality of preparation of students who will enter the secondary schools and perhaps, finally, the two-year college is obvious. Of course, the accruement of the benefits of increased teacher competence must ultimately be shared by the entire society.

Let us assess now the influence the two-year college teacher of mathematics may have on the prospective secondary teacher of mathematics. A large number of prospective secondary teachers of mathematics probably begin their study of mathematics in college at the algebra or trigonometry level; therefore, many of the concepts that they will teach as secondary teachers will be those they most recently encountered in some form at the two-year college level. If it is accepted that good mathematics teaching behavior by the prospective teacher can be fostered by modeling good mathematics instruction, then the importance of these first two years must be acknowledged. This component of the teachers' education may be as significant for changing or setting teaching behaviors as any other including the professional education component. Although the secondary teacher will continue his professional training in mathematics at the upper level, it would appear that two-year college teachers still have an unequalled opportunity to demonstrate the kinds of instruction that they would have these elementary and secondary teacher education candidates display in their classrooms for those students who ultimately may be students at the two-year college.

Again, as for the elementary teacher, the spirit of inquiry should be nurtured in the mathematics classroom, and the process rather than the product of mathematics should be the emphasis. Mathematics activities should involve more than mere repetition, exercise, or drill. Surely an understanding of the concept of the zeros of a function is much more important than the production of the roots of a quadratic equation through the use of a memorized formula. Unfortunately, somehow the relationship between these two things gets lost in a product oriented classroom. Clearly I am supporting an emphasis on conceptual learning and an emphasis on thinking. A psychologist with the Human Resources Research Organization has phrased it simply: "The new education must teach the individual how to classify and reclassify information, how to evaluate its veracity, how to change categories when necessary, how to move from the concrete to the abstract and back, how to look at the problems from a new direction—how to teach himself. Tomorrow's illiterate will not be the man who can't read; he will be the man who has not learned how to learn."[1]

Technological advances are now forcing us to examine our teaching objectives. Machines, relatively inexpensive ones, now do the things we in the past meticulously had our students do through algorithmic processes often routinely and mechanically performed. And the latter we called learning, and indeed I suppose it was, but is it the type that should receive priority today? As Toffler would put it, "Education must shift into future tense."[2]

The mathematical understanding of teachers must not be determined by those who still narrowly view the function of teachers of mathematics as that of teaching the students to produce the algorithms as they were taught to produce them. Professional monitoring must be performed by an informed, interested, sympathetic, and farsighted group. Should not those involved in the teaching of mathematics at least partially compose this group? Certainly no other group exists who should be more seriously involved in determining the course of mathematics teacher education, and no other group exists who can produce the needed improvements in teacher education. Higher education sorely needs people who know mathematics and who are interested in increasing the competence of those who teach it, yet this responsibility is too often being forfeited to those opportunists who feel that all the problems of the teaching profession can be solved with a carefully written behavioral objective. Perhaps the era of accountability will more prominently place involvement in teacher education activities among the list of criteria for promotion and tenure of mathematics faculties. Even at the present there is a growing list of college teachers of mathematics who are taking a more than "ivory tower" interest in the competence of teachers.

---

[1] Alvin Toffler, Future Shock, Bantam Books, 1970, p. 144.
[2] *Ibid.*, p. 247.

Recognizing the need for improvement in teacher education, specifically what can the two-year college teacher do to produce change beyond that directly associated with the classroom? The recent publication by the National Advisory Committee on Mathematics Education of the Conference Board of the Mathematical Sciences gives an excellent overview and analysis of elementary and secondary school mathematics and teacher education and provides background reading for those interested in reforms in teacher education. Some basic avenues of reform may be found in consideration of the following questions proposed for the two-year college mathematics teacher:

Do you confer with those involved in efforts at upgrading teacher certification standards?

Do you confer with colleges regarding their teacher education requirements?

Do you serve as consultants for elementary and secondary teacher workshops and inservice?

Do you keep current on the mathematics curricula of the elementary and secondary schools?

Do you read the literature of the teaching profession?

Do you participate in local, state and national organizations concerned with teacher preparation and improvement?

Do you offer on-campus or extension courses designed especially for the inservice teacher?

I do not mean to imply that all of the above are feasible activities for the two-year college mathematics teacher. They are only offered as the kinds of activities which can bring about change. I am not suggesting that the two-year college mathematics teacher should replace the competent people already working in teacher education. I am suggesting, however, that where improvements can be made or need to be made the two-year college faculty should work to that end.

PANEL: MATHEMATICS IN OUR CULTURE

Moderator: R. H. McDowell

THE FRESHMAN LIBERAL ARTS COURSE

Morris Kline

In his famous poem "Dover Beach" Matthew Arnold expressed his despair:

> "... for the world, which seems
> To lie before us like a land of dreams,
> So various, so beautiful, so new,
> Hath really neither joy, nor love, nor light,
> Nor certitude, nor peace, nor help for pain;
> And we are here as on a darkling plain
> Swept with the confused alarms of struggle and flight,
> Where ignorant armies clash by night."

Why was Arnold so despondent? He had just completed a freshman liberal arts course and he concluded that if mathematics, the noblest of man's creations, had no more to contribute to our culture than what he had been taught, then surely life had little to offer beyond the misery he was describing.

The greatest threat to the life of mathematics is posed by the mathematicians and their most potent weapon is their pedagogy. The best evidence for this assertion is the treatment of the freshman liberal arts course. Through this medium mathematicians have the opportunity to meet the greatest number of college students taking mathematics. Beyond numbers this group includes not only very bright students but the ones who will become leaders in our society and be in a position to support or influence the support of mathematics research and education. Hence teaching a suitable liberal arts course is more important than much of the current research and at least as important as training research mathematicians, most of whom will not do research. And yet it is in teaching just this course that mathematicians fail miserably.

What are liberal arts students taught? The most common courses given today offer set theory, symbolic logic, the theory of numbers, Boolean algebra, abstract algebra (the theory of structures), the axiomatic development of the real number system, and such trivial and peripheral topics as the Möbius band and the Königsberg bridge problem.

What is wrong with these topics? Set theory may perhaps be the foundation for the sophisticated approach to mathematics being developed by the Bourbaki school but it is of no use in understanding the mathematics that can or should be taught at the college level. Moreover, as usually taught set theory includes infinite sets. The concept of an infinite set baffled and was rejected by the best mathematicians until the 1870s and is still unacceptable to many today. Since the study of infinite sets does not return even finite riches to the student he does not see why he should attempt to make his reach exceed his grasp.

The theory of numbers deals among other matters with prime numbers. To a few mathematicians these are delightful, intriguing members of the number system. To the students they are hostile strangers. When they learn that there is an infinity of prime numbers they become convinced that the world is full of enemies. Ah! But the theory of numbers also teaches congruences, which, roughly speaking, teaches arithmetic such as our clocks utilize. And so students who have just about learned that $9 + 4 = 13$ are now taught that $9 + 4 = 1$. The mere mention of clock arithmetic makes students look anxiously at their watches to see how many more minutes must elapse before the period is over. Of course the theory of numbers does offer intellectual challenge and aesthetic satisfaction to sophisticated mathematicians, but students who have already been soured by eight years of arithmetic and two or three years of seemingly pointless algebra and geometry are not at all likely to respond to the values that professionals find in that subject.

In the highly artificial, logical development of the real number system students are taught among other topics the logical approach to negative numbers. The mere mention of negative numbers calls to mind the earlier teaching that negative numbers are used to represent temperatures below zero and thereafter the students' minds freeze. Irrational numbers approached logically are intellectual monsters, and for the first time students appreciate mathematical terminology. The entire development of the real number system on the basis of Peano's axioms is artificial, contrived, stultifying, useless and boring. Poor Gauss! He didn't know how to work with real numbers because he was born too soon.

Apropos of the logical development of the real number system many authors advance it as an example of how mathematics builds models for the solution of real problems. This is not the place to discuss applied mathematics but it is very clear that the authors who make such a statement about models haven't the least of idea how mathematics is applied. The example is absurd on many accounts. Let us note two. The real number system has been in use since about 3000 B.C., roughly 5000 years before the logical "model" was constructed. But no one could have used the real number system *model* before it existed. Secondly, the artificial, complex, logical construction is as far removed from reality as heaven from earth. No one would ever think of using it to predict anything even about real numbers, let alone physical applications. The objective in constructing the logical structure of the real number system has nothing to do with real problems. In the late nineteenth century mathematicians had reasons internal to mathematics to base every subject on a clear, explicit axiomatic basis, no matter how contrived the axioms had to be. Because this axiomatic basis had not previously been supplied for the real numbers several men proceeded to build it. Perhaps professors know all this and use the word "model" because it embraces other more desirable kinds of models and the word may suggest these more pleasurable kinds to the students. Unfortunately the logical model of the real number system lacks flesh and blood.

Symbolic logic, which presents the ordinary principles of reasoning in symbolic form, is a farce as an approach to teaching reasoning. To know what symbolism to use one must already know what the common meanings of "and", "or", "not", and "implies" are. But the students do not have these clearly in mind and symbolic logic conceals them under a maze of meaningless symbols. How ridiculous to teach symbolic logic to students who still confuse all A is B with all B is A.

Worse than that, professors teaching symbolic logic are courting trouble. As we all know $p \supset q$ is correct if p is false and q is false. A false proposition implies any proposition. Hence the assertion "If the moon is made of green cheese, then Gerald Ford was elected president," is a correct implication. Further, since $p \vee q$ is true if p is true or q is true and if a student says, "5 + 6 is 11 or 12," his answer is correct.

Boolean algebra, which is closely related to symbolic logic, is taught on the assumption that the liberal arts students are going to be electronic engineers. But the application to switching circuits suggests to the students that they switch courses.

The liberal arts courses purport to teach the power of mathematics and they do this by teaching abstract structures such as groups. What is done with groups evidences the power of mathematics as much as the study of philosophy shows how to run a spaceship.

What is the major problem facing our civilization? War? Inflation? Unemployment? Health? No. Judged by the liberal arts texts it is the Königsberg bridge problem. It is true that some 200 years ago the citizens of the village of Königsberg in East Prussia amused themselves by trying to cross seven nearby bridges in succession without recrossing any one. However, Leonhard Euler soon showed that the attempt was impossible. But mathematicians will not let the dead rest in peace and they revive the problem as though it were the most momentous one of our times. The villagers may have amused themselves in their walks on sunny afternoons but students are not amused to have the problem resuscitated in the chill of a gloomy mathematics classroom.

Still another favorite topic is axiomatics. The proper name for this topic is postulate piddling. My objection is best expressed in the words of Hans Lewy: "Too many people are making frames and not enough people are making pictures." Students are indeed surprised and gratified to learn that nine and one half axioms can replace ten.

No worse a collection of dull, remote, useless and sophisticated topics could have been chosen. They are not representative of mathematics or culture. Many come from the foundations of mathematics where only specialized and professional needs justified their creation. With a few exceptions they are developments that came long after most of the greatest mathematics we have was created. The best mathematicians of the past, Archimedes, Descartes, Newton, Leibniz, Euler, Lagrange, Laplace, Cauchy, and Gauss used almost none of them for the simple reason that they didn't exist. And even the great mathematicians of the present do not use most of them except in specialized foundational studies. The topics have about as much value for liberal arts students as learning to dig for clams has for people who live in a desert.

A common alternative to the above mélange of topics is a presentation of technical mathematics which starts about where the high school courses leave off and covers more advanced techniques. Some professors teach their own specialties, graph theory or group theory. In a liberal arts course there should be no technique for the sake of technique. This technical course is no better than the college algebra and trigonometry that the colleges used to require of *every* student. Such a course gives a low return on the students' investment. They are asked to surmount technical hurdles without being at all clear as to what the enterprise is all about. They are asked to perform mental athletics which leave them tired and dispirited instead of refreshed and stimulated. The technical courses offer brick-laying instead of architecture and color-mixing instead of painting.

Another common type of liberal arts course offers puzzles, curiosities, and trivialities. No one topic is pursued in depth and the topics are disconnected so that the professor can choose what strikes his fancy. The texts for this type are usually "enriched" with cartoons. Perhaps pointed, truly humorous cartoons can be admitted as a pedagogical device on the college level but sequences of drawings shallow in content and which hardly elicit a smile from a six-year old make no contribution.

Another alternative type of liberal arts course frequently offered and one which has acquired great vogue in the last fifteen or so years is generally known as Finite Mathematics. Just what is finite is not at all clear unless it be the students' attention span to it. It does not include any calculus but it does use real numbers and complex numbers and algebraic processes and theorems which involve infinity in several ways. The content, like that of the typical liberal arts course, is a conglomeration of topics having little relationship to each other. Such a course, if the topics are properly chosen, might be useful to social science students. Presumably then it would contain applications to the social sciences. On examination one finds a mathematical system which describes the marriage rules of some primitive Polynesian society. Finite Mathematics is a fad if not a fraud. In any case it is not a liberal arts course.

Many would argue that the contents of a liberal arts course is not the chief criterion of its value. The main objective is to teach students precise reasoning. A course in mathematics proper may teach sharper reasoning but the students have already had three or so years of mathematics in high school, and it would seem that whatever training of the mind mathematics can supply would have been supplied already. Actually the vaunted value of deductive reasoning is grossly exaggerated. The most important problems of life are not decided by deductive reasoning. Judgment, the weighing of evidence, inductive reasoning, and reasoning by analogy are far more vital. Some such thinking could be taught in mathematics but deductive reasoning is the only one featured. In any case the question of whether learning to think about mathematical themes improves thinking in other spheres remains open. Certainly whatever faculties equip a man to understand and judge wisely about human problems are not more widely found among mathematicians. The distinctions that must be made in analyzing character, personality, values, and good and bad behavior are far more subtle and call for a more highly perceptive and critical faculty than anything mathematics will ever teach. Deductive reasoning is not the paradigm for the life of reason.

When this defense of mathematics is attacked the professors fall back on the aesthetic satisfactions mathematics offers. All the preaching and rhapsodizing about the beauty of mathematics will not make such ugly ducklings as the logic of the real number system more appealing. There are beautiful portions of mathematics but the attempt to sell the beauty of mathematics to the liberal arts student is doomed to failure. Beyond the point already made that students are soured by their elementary and high school experiences, there is the obstacle that the beauty is esoteric.

Many professors believe that the goal of a liberal arts course should be to teach what mathematicians do and that this is what they are attempting. No more effective means of driving students away from mathematics have ever been devised. What do mathematicians do? They strive for personal success and even neglect the interests of the very students they say they want to attract. But mathematicians create. Do these courses then teach the fumbling, the guessing, the blundering, the testing of hypotheses, the false proofs, and other acts of the creative process? No. They teach theorem and proof as though God inspired the mathematicians to proceed directly to the finished product.

Mathematicians are narcissists. That's tolerable. But they have made mathematics part of themselves and so they offer topics *they* value in the vain expectation that students will value them. But mathematics *proper* has very little to contribute to a liberal arts education. Those who believe that such a statement is too strong should heed the words of Hermann Weyl: "Mathematics is not a natural concern of man. It has the inhuman quality of starlight, brilliant and sharp but cold. But it is an irony of creation that man has been most successful where knowledge matters least—in mathematics, especially in number theory." Moritz Pasch said that mathemathical thought even runs counter to human nature.

What should a college course addressed to liberal arts students offer? The answer is contained in the question. The liberal arts values of mathematics lie in the contributions which mathematics has made, directly and indirectly, to our understanding of the physical world, to technology, to philosophy, especially the problem of truth, painting, music, political thought, economics, religious thought, and literature. In short, the liberal arts course should emphasize the role of mathematics in our culture. This is the type of course which a non-user of mathematics would be far more willing and able to appreciate and be a genuine contribution to his education.

The richness and value of mathematics derive primarily from its use in studying the real world. The concepts and reasoning of mathematics serve to obtain results about physical and social phenomena. Mathematics is a means to an end. It may be unfair to compare mathematics with a hammer. But the exaggeration may make the point clearer. One could study hammers in and for themselves but hardly anyone would see much point in this. The hammer as a tool, however, is not only effective but indispensable. Even the most primitive peoples made hammers out of stones and wood.

Max M. Schiffer, professor of mathematics at Stanford University, has pointed out that, "The miracle of mathematics is that paper work can be related to the world we live in. With pen or pencil we can hitch a pair of scales to a star and weigh the moon. Such possibilities give applied mathematics its vital fascination. Can any subject give the would-be mathematician—initially at least—a stronger and more natural interest? And what about the non-mathematician? Deny him introduction to this subject, and his appreciation of our cultural heritage must inevitably be inadequate. For mathematics in the broadest sense is instrumental not only to our understanding, but also to our changing the world we live in." Mathematics proper may be a monument to human inventiveness and ingenuity but it is not in itself an insight into reality. Insofar as it helps us to secure that insight it is important. This then is what we must teach.

The prime accomplishment of a truly liberal arts course should not then be mastery of mathematics proper but an appreciation of the role of mathematics in Western culture. Appreciation rather than skill has long been recognized as an objective in literature, art and music. It is equally justifiable as an objective in mathematics. Of course to teach the cultural values we must teach the mathematics that is involved. But we cannot stop short with just the mathematics. When students can appreciate why Matthew Arnold wrote the lines I quoted at the outset, why they speak trigonometry every time they say a word, and why the founders of our government began the Declaration of Independence with the words, "We hold these truths to be self-evident ..." then we shall have succeeded in teaching a liberal arts course.*

Knowledge is a whole and mathematics is part of that whole. However, the whole is not the sum of its parts. The present procedure in the liberal arts course is to teach mathematics as a subject unto itself and somehow expect the student who takes only one college course in the subject to see its importance and significance for the general body of knowledge. This is like giving him an incomplete set of pieces of a jig-saw puzzle and expecting him to put the puzzle together. Liberal arts mathematics must be taught in the context of human knowledge and culture. Mathematics has played and continues to play a central role in the fashioning of Western civilization. It is therefore one of the fundamental obligations of a responsible mathematics curriculum to present this value. The professors who do not do so have shortchanged the students.

Professors must learn that mathematics proper is not the most important subject for the non-professional. Even some of the best professional mathematicians did not grant the subject supreme importance. Newton regarded religion as far more vital and said that he could justify much of the drudgery in his scientific work only on the ground that it served to reveal God's handiwork. But of course Newton was just a lowly physicist. Gauss ranked ethics and religion above mathematics, but Gauss, too, devoted most of his life to physics and astronomy.

The elitist, narcissistic mathematician who presents his own values, beauty, curiosities, and trick problems, is totally unfit to be a teacher in a liberal course. He is truly culturally deprived. He is as ignorant as a citizen of our democracy who is unaware of our constitution and the political principles which it lays down.

The gap between the sciences and the humanities is often blamed on the humanists on the ground that they do not posses the will or the intelligence to learn the sciences. But the mathematicians are not only unwilling to learn the humanities, they are also unwilling to meet their obligation to teach a humanistically oriented course in mathematics. It is ironic that professors

---

*I realize that in an article one cannot do justice to the subject of the proper liberal arts course. With much concern as to misinterpretation of the motives I mention my *Mathematics: A Cultural Approach* (Addison-Wesley, 1962) and my *Mathematics for Liberal Arts* (Addison-Wesley, 1967) as fuller expositions of versions of a liberal arts course. However I can readily conceive of fine liberal arts courses which would differ considerably from these two versions.

teaching in a liberal arts college, which is purportedly devoted to educating the whole man and to instilling interests and attitudes, are not themselves interested in learning material closely related to their own subject. If the contents of the various liberal arts courses are any indication mathematicians are merely proficient specialists and specialization is the biological niche for mediocre minds.

The fashioning and teaching of a suitable liberal arts course is not just a matter of fulfilling our obligation to students. Mathematics is disappearing as a requirement for a college degree. At best it is now an option which students choose in preference to a science course which calls for laboratory hours. And the few who do take and complete the course are grateful only for the fact that it is over and vow never to become involved with mathematics again. If we can't do better or won't do better let us drop the farces we offer as a liberal arts course and let us stop wasting students' time. In fact if we don't, students will stop us.

We talk much these days about the public image of mathematics. In the liberal arts course we have the best opportunity to reach a large group and present a favorable public image. Let us use it and serve ourselves and the students. May I close with a quotation from Plato: "Now, when all these studies reach the point of inter-communion and connection with one another, and come to be considered in their mutal affinities, then, I think, but not until then, will the pursuit of them have a value for our objects; otherwise there is no profit in them."

THE VICIOUS VERSUS

R. A. Rosenbaum

> How much has happened in these fifty years—a period more remarkable than any, I will continue to say, in the annals of mankind. I am not thinking of the rise and fall of Empires, the change of dynasties, the establishment of governments. I am thinking of those revolutions of science which have had much more effect than any political causes, which have changed the position and prospects of mankind more than all the conquests and all the codes, and all the legislators that ever lived.
>
> Benjamin Disraeli, 1873

> I cannot help thinking that it would be a distinct gain to the ethical as well as to the intellectual standing of the clergy, if every man who enters the ministry had done some considerable amount of laboratory work in some department of science, so as to acquire the power of exact observation and absolutely truthful description, and had associated with scientific workers sufficiently to feel the influence of the scientific habit in cultivating the sense of veracity.
>
> William North Rice, 1908

How wonderful was that optimistic era when mathematics and science were thought to hold the key to a golden future! The technology which was to develop from science would ensure peace and plenty; in his ample leisure, every layman would enjoy the enlightening lectures of the successors of Thomas Henry Huxley; mathematicians had merely the task of realizing Hilbert's goal, and scientists, of tidying up the edges of an almost complete physics and a well-developed biology; and even ethics would become simple and straightforward when subjected to the disciplines of logic and the scientific method.

But how do we fare now? We gag on water polluted by detergents and by chemicals suspected of being carcinogens; we choke in ever diffusing smog; we curse the automobile fender made of tin-plated oatmeal; we tremble when we think of the bomb and of population increases; we shudder when

we try to cope with automation, with energy shortages, and with unman-
ageable urban problems. When C. P. Snow calls on the present-day Huxleys
to explain the second law of thermodynamics to the man in the street, he
gets no answer, for the scientists are too deeply immersed in their specialties to
have time for such teaching; and J. Robert Oppenheimer says that it makes no
difference anyhow, because contemporary physics is too difficult for anybody
but an expert to understand. Problems that once seemed simple have grown in
complexity, and the mathematician and the scientist experience the frustrations of Sisyphus. Then comes the cruelest twist of all—science, far from
smoothing the way for ethics, itself raises a host of ethical problems, tangled
and subtle.

Small wonder, then, that we live in an age of pessimism and anxiety. We
seem to have created science and technology only to become their slaves.
Like most slaves, we are not treated too badly by our masters in relatively
unimportant matters, but we have little control in major issues. Oh, to return to the simple days, the simple ways! Some restore their optimism and
banish their anxiety by a procedure frankly irrational; they pretend that
science doesn't exist. The TV tube and the refrigerator and the hand-held
calculator are white magic for which all know the incantations; the transis-
tor and the antibiotic and the calculus are stronger magic for professional
conjurers; the meson and DNA and topological dynamics are only for a few
sorcerers.

Ignorance and, perhaps, fear of the effects of science have led many people
to believe that mathematics and science are cold, mechanistic activities—at
best, inhumane; actually, probably *inhuman*—surely less worthly than the
humanities of the attention of thoughtful, sensitive individuals. This is the
thesis advanced by the philosopher Brand Blanshard in an article entitled
"Hamlet Versus the Second Law of Thermodynamics" in the N.Y. *Times
Magazine* some years ago. Mr. Blanshard wrote, "I admire at a great distance the skill of the mathematician in manipulating his symbols according
to his recondite rules, just as I admire the astonishing gift of young Bobby
Fischer for manipulating the men on a chess board."

An equally distorted view is held by the movie-maker, Frank Capra, who
is quoted as saying, "Math and logic and creativity don't go together." The
same attitude is expressed by the novelist Norman Douglas, who has one of
his characters in *South Wind* say, "Mathematics ... a medieval halo clings
round this subject which, as a training for the mind, has no more value than
whist-playing ... As a training in intelligence it is harmful; it teaches a person to underestimate the value of evidence based on their other modes of
ratiocination. It is the poorest form of mental exercise—sheer verification;
conjecture and observation are ruled out ... If you mention the utility of a
mathematician like Isaac Newton, don't forget that it was his preeminently
anti-mathematical gift for drawing conclusions from analogy which made
him what he was."

And finally, the poet, Paul Valéry, one who himself had a genuine appreciation of mathematics, expresses the popular view: "The idea of Poetry is often contrasted with that of thought and particularly Abstract Thought. People say, 'Poetry and Abstract Thought' as they say Good and Evil, Vice and Virtue, Hot and Cold."

I suspect that many mathematicians, whose social intercourse is restricted to a circle of other mathematicians, to a few non-mathematicians on their best behavior, and to a group of dedicated or intimidated students, may believe the foregoing quotations to be vastly overdrawn. Let me state as forcefully as I can: candid statements about mathematics made by the average shopkeeper, or member of Congress, or doctor, or housewife, or engineer, or plumber, or industrial manager, or grade-school teacher, or lawyer, or secretary would be even more pungent and pejorative, even more curdling and caustic, than these.

This is a bad situation. It is bad for mathematicians as members of our society; it is bad for mathematics as an element of our culture; but, most important of all, *it is bad for our culture.*

It is probably true that the nature of mathematics and its role in our culture have *never* been well understood, even by the educated portion of society. The misunderstandings seem to be even worse these days than in earlier times, despite the expansion of formal education. In a parochial sense, we as mathematicians may be upset by the misunderstandings; but the truly serious issue relates to the future of a democratic society which must rely heavily on technology and science, and, ultimately, on mathematics. It is far from healthy for a society to rely for its survival on the activities of a secret priesthood or a group of sorcerers.

How did we come to our contemporary plight, in which the general understanding and appreciation of mathematics stand at such a low level? Like Professor Kline, I lay most of the blame on our abysmal teaching record; but there are other factors, too, like the widespread flight from reason, which seems to have accelerated in recent years. One may well ask, "What causes a flight from rationality? What is cause and what is effect?"

I shall not try to answer these last questions, for they are largely irrelevant to our course of action. As mathematicians and teachers, with a concern for our subject and a sense of responsiblity to our society, it behooves us to exert every effort to bring as many of our fellow-citizens as possible to *appreciate* mathematics—literally, to *understand* it, to *enjoy* it, and to *esteem* it. I go so far as to suggest that, for many of us, such an effort is likely to have greater social usefulness than anything else we do. Articles for the popular and semi-popular press, preparation of teachers in elementary and secondary schools, involvement in interdisciplinary "general education" courses and in adult or continuing education programs, participation in "community leadership" and parent-teacher forums—these are some ways to attack the problem.

As members of the Association, most of us will feel that the place for us to start in ameliorating the situation is in our collegiate teaching. With some of Professor Kline's criticisms and suggestions, I am in whole-hearted agreement; but I go farther than he in one important respect: I believe that it is fruitful to bring *mathematics*, and not only discussion about mathematics, to non-science undergraduates.

For example, the topic of non-Euclidean geometry has many features of pedagogical advantage:

In the first place, the very notion that there might be *more* than one parallel is itself so shocking as to capture immediately the interest of all but the comatose.

Next, the attempts to prove the Euclidean parallel postulate illustrate "mathematics in the making", remind us that mathematics is created by human beings, and emphasize that there are reasons in addition to a desire for economy and elegance for examining an axiom system.

Also, the virtually simultaneous development of non-Euclidean geometry by several people exemplifies a common phenomenon, the psychological and sociological aspects of which are interesting to consider.

Moreover, models of non-Euclidean systems and the usefulness of non-Euclidean geometries in modern physics link the abstractions of mathematics to the "real world".

Finally, the impact of the development of non-Euclidean geometry on the mathematics of the past hundred years, and also on epistemology, can be seen to have been revolutionary, with repercussions still felt. Students can appreciate various *significances* in the subject.

In my experience, students are excited by this topic, discuss its implications in other courses, and ask, "How come nobody ever told me about this before?"

Better late than never.

It is not my intention to list topics suitable for a course for non-science students, but I cannot resist mentioning another—that of instantaneous rate of change. Here is an idea of far-reaching importance, with diverse examples, in which the *generality* of the method of solution should prove appealing even to naive students. But more than this—the *conceptual* difference between the average rate of change over a small interval and the instantaneous rate of change can be made clear through discussion of the Newton-Berkeley controversy, illustrating one aspect of what Dirk Struik has called "man's struggle with the infinite", and thus helping the student to grasp a notion of central importance in the development of mathematics.

In introducing my examples of topics for non-science students, I spoke of "mathematics, and not only discussion *about* mathematics." But in my elaboration, I have made it clear that I favor considerable "discussion *about* mathematics, and not only mathematics" in such a course. It is essential, I think, explicitly to present both *mathematics* and its *setting*: its history, its practitioners, its applications, its significance.

And, whatever we do, we must modify our presentations so that mathematics will not be viewed as "the poorest form of mental exercise—sheer verification; conjecture and observation are ruled out." For years, George Pólya has been urging us, "Let us teach guessing", and we have been acknowledging the validity of his advice. Nonetheless, and despite the magnificent examples which Professor Pólya himself has set before us, we persist in presenting mathematics only in retrospect, as a completed structure, devoid of the struggles, the intuitions, the frustrations, the hunches, the despairs, the insights, the glorious rewards of long, hard work, all of which make mathematics a fascinating *human* endeavor.

We tend not only to be silent on this aspect of mathematics but to shield our students from having any experiences which would lead them to discover for themselves the fascination which we find in our work. No wonder, then, that George Garrett writes in one of his short stories, "He began to conceive of [a certain individual] as a kind of geometric figure, bloodless but well-made." Surely, an ellipse is bloodless, but so is a fugue. The imaginations inspired, the feelings evoked, in a full-blooded human being, by a contemplation of the properties of an ellipse may be as complex, as ramified, as those induced by listening to a fugue. It takes innate sensitivity and considerable education to get the most out of either.

Similarly, as Hermann Weyl has said, starlight *is* brilliant, sharp, cold, literally with an inhuman quality. But the questions that have been stimulated in the minds of human beings by a consideration of starlight have led to some of the greatest achievements of the human race—achievements which possess transcendent beauty and practical significance—achievements which have helped to lift mankind beyond the concerns of mere daily existence.

It appears to be the *vitality* of great literature, in contrast with the "bloodlessness" of mathematics and science, that causes Brand Blanshard to set up the dichotomy of "Hamlet *Versus* The Second Law of Thermodynamics". But the dichotomy ignores the fact that the grand problems of mathematics and science are formulations of profound human questions—not universally felt, not even widely articulated—but nonetheless inherent in man's quest for understanding and wisdom.

We agree with Mr. Blanshard that "a man is not a whole man, he is a maimed and stunted man, if he is blind to Rodin, deaf to Mozart, indifferent to what Eliot, Gropius, and Picasso have been trying to do." Will he not agree with us that a man is also not a whole man, he is a maimed and stunted man, if he is unaware of Galois, ignorant of Gauss, insensitive to what Cantor, Poincaré, and Gödel have been trying to do?

Rather than take sides on Mr. Blanshard's *versus*, I prefer to follow Emerson who expresses an old-fashioned view:

"We do not listen with the best regard to the verses of a man who is only a poet, nor to his problems if he is only an algebraist: but if a man

is at once acquainted with the geometric foundation of things and with
their festal splendor, his poetry is exact and his arithmetic musical."

Mr. Blanshard ends his article with two cheers for the humanities, perhaps
leaving one cheer for mathematics and the sciences. I wouldn't be any hap-
pier with two cheers for mathematics and one for the humanities, or with
one-and-a-half for each.

I think that it is important for our next 200 years that we work hard to
get him *and lots of others* to join all mathematicians in giving three hearty
cheers for both!

PANEL: THE TEACHING OF MATHEMATICS IN COLLEGE: A 1976 PERSPECTIVE FOR THE FUTURE

Moderator: C. V. Newsom

COMMENTS MADE WHEN INTRODUCING THE PANEL DISCUSSION

C. V. Newsom

"The Teaching of Mathematics in College: A 1976 Perspective for the Future," the subject of the next panel discussion, is receiving our attention at a time when a unique atmosphere of concern exists within the academic world. "In the brief span of about five years," Freeman and Hollomon* of the MIT Center for Policy Alternatives assert, "the college job market has gone from a major boom to a major bust." The severe turn-around, as Freeman and Hollomon see it, "is a far-reaching unprecedented development of sizeable dimensions." A complex of factors provides the basis for the disturbing phenomenon: The depressed economic state of the nation has produced extensive unemployment of people irrespective of their educational attainments but, in addition and of major significance for us, there has been a pronounced erosion of the traditional belief held by the American people that college graduates are special people who are capable of distinctive accomplishments within the framework of American life. It is expected, as Freeman and Hollomon have indicated, that the sudden decrease in demand for college graduates will have an early and significant effect on college enrollments, on institutional budgets, on curricula, on the nature and size of college faculties, and on the strategy of teaching. Departments of mathematics in college, as is true of virtually all academic departments, are concerned, and properly so, by the disconcerting trends.

In fact, the year 1976 must be acknowledged as a year with distinct challenges for those who have major responsibility for determining the nature of the mathematics program in our colleges and universities. The panel presentation which follows, conducted by two of the best known

---

*Richard Freeman and J. Hollomon, "The Declining Value of College Going", *The American Future*, fourth in a series by the Center for Policy Alternatives, MIT, Cambridge, Mass.

members of our mathematical community, is concerned with the most important of those challenges. Israel Herstein, the first panelist, will suggest some answers to the question, What will be expected of the good college mathematics teacher? Peter Hilton, second panelist, will deal with the question, What experiences should be provided in Graduate School to prepare the college mathematics teacher? Both men have received wide recognition, both in this country and abroad, for the excellence and the versatility of their mathematical accomplishments; if I told you in any detail of their accomplishments, it would require a substantial part of this hour. Look them up in *Who's Who*. Both men are known as unusually able expositors of mathematics and as exponents of good teaching.

WHAT WILL BE EXPECTED OF THE GOOD COLLEGE MATHEMATICS TEACHER?

I. N. Herstein

Let me begin by being properly literary, with a quotation from the poem "Locksley Hall" by Alfred Lord Tennyson:

> "For I dipt into the future, far as human eye could see,
> Saw the Vision of the world, and all the wonder that would be."

In all honesty, I always found Tennyson to be an insipid poet, and the lines I quoted particularly trite. However, today I shall dip into the future, albeit not as far as human eye can see. I'm afraid that many of you may find what I see there as neither wondrous nor wonderful.

I am supposed to speak on "What will be expected of the good mathematics teacher in the future?" Now, prophecy is not exactly my line of business, and as a prophet I'm usually a total loss. Nevertheless, I will try to describe the situation, as I see it, for the very near future. What I shall say is full of my own biases and prejudices; there is no doubt that the difficult job situation for our young mathematicians has played a significant role in shaping my thoughts on these matters. As will become abundantly clear in a few minutes, I have some particular axes to grind, and grind them I shall.

To my mind three fundamental variables play a key role in determining the functions and nature of the mathematics teacher at the college level in the next few years. The first of these is: what will be the nature of mathematics itself in the near future? The second is: to whom will we be doing the lion's share of our mathematics teaching? The third is: what will we be teaching them?

I shall address myself to these questions in turn. Interspersed with what I will have to say will be some suggested roads to follow to allow our future teachers to be more adequately prepared for their tasks. In this I may encroach on the territory which has been staked out for Peter Hilton. I hope he will forgive my intrusions.

What seems manifestly obvious to me (and to many other people) is that the hey-day of the highly abstract is, for the moment, over in mathematics. What is taking place is a concretization, a focusing on particular problems, a detailed development of a variety of mathematical systems. And our field is not unique in this. To cite one example, painting; it is undergoing a similar pattern. We see now a return to realism, a sharpened interest in surrealism. Even in the paintings of Wilhelm de Kooning, a leading exponent of the Abstract Expressionists, we now see human figures and a strongly realistic element. But this is not a return to the stilted style of realism of the Pre-Raphaelites or the Barbizon school. Instead it is a creation of a new type of realism, of a much looser structure. I expect the same to happen in mathematics; not a return to realism, perhaps, but a return to reality.

I'll probably be accused of advocating narrowness or parochialism, so let me be parochial and use examples and experiences from my own area of interest, algebra. I'm sure that analogs of these exist in all areas of mathematics.

I recently heard two lectures by my distinguished colleague Saunders Mac Lane on the rise and decline of abstract algebra. If I understood him correctly, what he said was that abstract algebra rose and flourished as a *movement* from 1921 to 1971, at which point it began to decline. Not that abstract algebra is now dead, but it has assumed a new form and is no longer a movement. A retrenchment has set in, and with it a desire to apply all the highly abstract gadgetry developed in the preceding period to more concrete situations.

Let's take an example. In my opinion, the best and most exciting work being done in the abstract algebra today is in the theory of finite groups, especially in that part pertaining to the classification of the finite simple groups. This is highly concrete work, on a highly specific category of objects, using long, hard, often dirty arguments. But there is no argument with the spectacular successes these people have achieved. I see a similar phenomenon—a concentration on the particular—in ring theory, in Lie algebras, and in most parts of algebra with which I have some acquaintance.

I find this trend towards the specific a healthy one. The books written by Bourbaki are magnificent and have played a crucial, central role in the development of several generations of mathematicians. However, I feel that their basic philosophy that mathematics is a unified whole, or perhaps more accurately, the effect of this philosophy on many young mathematicians, has been deleterious. There are some mathematicians—damn few, I might add—who have either the talent or the perspective or both to scale a lofty peak and see the whole mathematical panorama stretched out below them. Our field desperately needs such people; they are often the ground-breakers and direction-finders for new areas of research. But most of us are lucky if we can drag ourselves up to the top of a little knoll and manage to see, from this vantage point, a small fertile patch.

In other words, I'm arguing that most of us are best off concentrating our efforts on specific problems, in specific areas; that for a large number of us our major contribution should be that of problem solvers. These problems can be in mathematics itself, or in any other field of endeavor—wherever they arise. But let the problems be good!

I don't believe that this contradicts my basic conviction that our students must get a wide, thorough education in the basics of the major mathematical fields. To solve a good problem requires a broad, deep understanding of mathematics, and a richness and variety of its techniques. It also implies a need to know enough about other areas of work, outside of mathematics *per se*, to be able to communicate effectively with the people in these fields for whom these problems constitute serious obstacles.

What has this to do with the teaching of mathematics? Plenty! For, if the students we train will have as one of their principal roles that of problem solver and of mathematical service to others—and I'm convinced that this will be the case—we better have teachers who can teach their students to solve problems, to be more flexible in their view of their responsibilities, to be less pristine and aloof.

How can we achieve this? First of all, we must teach our students that mathematics does not exist in a vacuum, that by some fortunate quirk of nature, mathematics has a role to play outside of its own immediate context. Surprise, surprise, but it can be used and applied. And it isn't only elementary mathematics, or humdrum mathematics that finds application elsewhere. Let me cite three examples—out of a very rich sample of possibilities—from algebra. One of these is from things done 20–30 years ago, the other two from work going on right now.

The first, and oldest, example is the use of the very beautiful and powerful results of Frobenius on matrices with non-negative entries, in economics. Such application can be found in the work of Metzler, Solow and others. This is a form of problem solving, the problems arising in the need to set up and to analyze a model of a specific set of economic situations. And here we see, to reach the ultimate goal of answers, a genuine application in an allied field of non-trivial algebraic facts to derive non-obvious results.

The second example is the application of the results of one part of mathematics—commutative ring theory—to solve specific problems in an other area of mathematics—combinatorics. This can be found in the work of Richard Stanley. He uses, highly effectively, very fancy, sophisticated, even esoteric results from commutative algebra theory to the solution (and by solution here I mean a definite answer) to a series of combinatoric problems that could be explained to the average individual. This is *applied* algebra at its highest level. Problem solving, but really good problem solving.

The third example is of a slightly different nature. It involves the description, in highly specific and concrete terms, of the simple objects in an algebraic theory. The area is that of classifying the finite-dimensional simple graded Lie algebras—work that is going on right now. The problems arose and were formulated by physicists in their study of supersymmetries. This sparked an interest among some outstanding American and Russian algebraists to tackle and to solve the problem, which they essentially have now done. The physicists may now have the tools to finish their job in this direction of their research. The nature of the mathematical problem, and the techniques for its solution, are alien to the physicist's way of thinking. It needed first-rate algebraists to carry it through. I might add, parenthetically, that the mathematics so developed, as an entity all by itself, is interesting mathematics of the highest calibre. So there has even been a pay-off to mathematics itself. But, I would still describe what was done as problem solving, admittedly at a somewhat rarefied level.

To summarize what I said earlier, I feel that at the undergraduate level we must teach our students the importance of mathematics as a tool and as a problem solver. We must give them the wherewithal to use this tool; we must teach them to keep their feet on the ground. Sure, they must learn a lot of abstract mathematics, but wherever possible they must be shown that these abstractions have something to do with a reality within or without mathematics.

To do so puts heavy demands on both the student and the teacher. But I'm going to insist on more. For the average mathematicians it will not be enough just to have a strong command of their own subject. For them to be able to practice mathematics as a profession, they will have to know other areas almost as well as they know mathematics. This requires time and effort.

What I urge is that our mathematics majors, from their very first days in college, carry a dual major. This implies that they specialize from their freshman year on in mathematics and X, that their training in X be as complete and thorough as that in mathematics. Clearly something will have to be given up. That something is the fulfillment of liberal arts credits and courses. I know that what I'm saying horrifies Educators who will claim that we will turn out highly trained technicians, culturally retarded, imbeciles outside of their own field. However, I feel it is imbecilic to expect that the thin patina of culture deposited on our students by a few liberal arts courses makes them culturally advanced or intellectually oriented. I don't believe, nor have I ever believed, that these courses have broadened our students. Nor have I seen a notable lack of culture or intellectuality amongst the mathematicians I know who have been trained in such kinds of dual major programs. In fact, in a large part of the world, university education is precisely along such lines. Naturally I would like the mathematician to be a broad cultured person. But culture is acquired through reading, through experiences and not by the token exposure to a few courses.

Be that as it may, I feel that this dual major is important, even necessary, for our mathematics students and prospective mathematics teachers. To use an odious term, their talents will be more "saleable", for, clearly, they will have more to offer. In addition, taking into account the kind of students they will be teaching, they will be better equipped to understand their students and to teach these students what they need, in a framework that these students can understand and appreciate.

I now turn to the second and third variables "whom and what will we be teaching?" Here, too, a definite pattern has been emerging in the last few years. Let me speak about what has been taking place at the University of Chicago, admittedly a highly atypical example. We have relatively few undergraduates—about 2500 out of a total student population of 7500—we have no engineering school and no undergraduate school of education. Yet, judging from what I have heard from friends at a large cross-section of colleges and universities, what has been happening with us is happening with them. In the past few years both the number and diversity of the students we face and teach have jumped markedly. In the past three years the number of students taught in our department at Chicago has gone up by 25%. With all this, the number of mathematics majors has decreased or, at best, stayed constant. We are teaching students from a wider and wider sector of the university. More and more outside departments insist that their students learn mathematics. I feel that we are seeing only the beginning of a trend which will become more and more widespread.

If this is the case then we obviously must take a fresh look at what our responsibilities will be in the university and in the community at large. One thing is clear to me. We no longer can afford to be embroiled in the kind of hassles we have had in the past with the engineering schools about the content and emphasis of our courses designed for their students. While we might have been able to fight off the engineers, we certainly can't take on the whole university. Nor should we want to.

If the statistics I quoted earlier reflect the present-day realities, then we are facing a very wide gamut of students. These students have ranging interests. Perhaps the only feature they have in common is that, while they need mathematics, they are not interested in the mathematics for its own sake. And we must take cognizance of this. We—that is the mathematics teachers of today and tomorrow—cannot blithely say (as we so often did in the past): here, we present you with mathematics as we see it, take it this way and, if you don't like it, the hell with you. It just isn't going to work.

We must be flexible, accomodating, even responsive, not because it is the politic thing to do but because it is the right thing to do. These students will be with us a certain length of time. It is our responsibility to use this time to make them as proficient and capable as possible with the mathematical tools they will need.

Don't get me wrong. I'm not suggesting that we compromise our mathematical principles, nor that we teach mathematics as a cookbook subject. Far from it. But I am suggesting that we slant our teaching and the course content more to the particular needs and talents of the students involved. Most important, that we view these needs and talents sympathetically.

In order to do this the teacher, or, at least, the set of teachers, must know enough about the disparate fields from which the students come, and about the way of thinking in these fields. Only this way can we set up meaningful syllabi for such courses. In fact, I would prefer that these courses have definite goals but somewhat amorphous syllabi. The good teachers, confident in their mastery of a block of outside material, would play everything by ear, tailoring the courses according to the student make-up of the particular sections.

Even in our purest courses, for our most abstractly oriented math majors—many of them, after all, will be mathematics teachers—we should constantly be showing the students how the course material is used. These math majors would be better off if we taught them somwhat less material and somewhat more about what this material means and can do. For instance, in teaching linear algebra why not digress to discuss some of the applications as in Noble's book, or to the treatment of economic models in some of the work I cited earlier? In discussing finite fields, why not develop algebraic coding theory, as one of the consequences of the general theory? This means that certain other material has to be left out. Fine. We can always pick up Galois theory, say, in the first year graduate course.

I recognize that it is difficult for us, with the kind of training, emphasis and point of view that we have had, to develop such material, devise such curricula, and write appropriate texts. But at least we can lay the ground work so that our successors—our students of today and their students— will be adequately equipped to carry through such programs.

Until now, the bulk of mathematical material that we have been teaching to students from outside fields has been at the calculus or pre-calculus level. I don't think that this will persist very far into the future. I foresee a widening and deepening of the subject areas that we shall teach others. As the outside fields become more sophisticated mathematically— and this is constantly taking place— they will demand that their students learn more, and more sophisticated, mathematics. To give a small example, in the most recent undergraduate course in abstract algebra that I gave, half the students were undergraduates from other departments, and the best student was an economics major.

The need by others of more sophisticated mathematics will accentuate our problems. While it may be easy to show the usefulness of the calculus in a large variety of situations, it will be much harder with the more esoteric material we will be teaching. And let's not forget the computer. Its glamor and ubiquity must be reflected in our offerings, not just in some *pro forma* way, but genuinely integrated into the mathematics taught and in the applications. It goes without saying that for our students to be effective teachers they will have to know a great deal about programming and theoretical computing. We should insist that all our math majors learn these things now.

Running through everything I have said has been the theme that our students and prospective teachers be more familiar with, and more at home with, the uses and applications of mathematics, that we and they get off our high horse about the service role of mathematics. Let's not forget that many of our mathematical heroes made substantial contributions outside of mathematics itself. Many astronomers and physicists consider Gauss as one of them. There are some fundamental results in particle physics that are due to Emmy Noether. A field created by von Neumann is now more properly a part of economics than of mathematics. And even G. H. Hardy, who revelled in the uselessness of mathematics, has a result in genetics named after him.

WHAT EXPERIENCES SHOULD BE PROVIDED IN GRADUATE SCHOOL TO PREPARE THE COLLEGE MATHEMATICS TEACHER?

Peter J. Hilton

First, let me commend to you the remarks made by my friend Professor I. N. Herstein. I entirely agree with the principles which he has enunciated, and I am very happy to have learned from him of examples which strengthen my own convictions about the relationship between pure and applied mathematics. I see my own role on this panel as that of providing some practical guidance as to how the objectives which we both desire may be achieved.

To answer the question which forms the title of my own contribution, we must first try to answer the related question, what do we expect that the college mathematics teacher will be teaching and whom will he be teaching. His students will be divided into four overlapping categories: (a) future users of mathematics, (b) future educated citizens, (c) future teachers, (d) future mathematicians. I take it to be common ground that he will be preeminently concerned with teaching mathematics to members of the first three of these categories, so I would like to concentrate on this aspect of the question. (In any case the problem of teaching future mathematicians is in my view essentially solved—many would say that it has been solved far too successfully!) I will claim that the problem of teaching the future educated citizen is a proper subset of the problem of teaching the future user of mathematics. I will also very shortly propound a proposition which will effectively allow me to set on one side the question of the teaching of future teachers, so that I feel justified in concentrating, at this stage, on the problem of teaching the future user of mathematics.

By the term 'user of mathematics' I intend somebody who uses mathematics in his work. He may therefore be an academic who works in some theoretical science. However, such a user of mathematics will, I believe, be rare, in the foreseeable future, compared with somebody who uses mathematics outside the academy; and therefore I wish to make the latter my paradigm. Thus, it seems that we should agree at the outset that the college mathematics teacher will be teaching with a view to his students having a good understanding of the nature of mathematics and of how it is applied to the real world. I will describe such a student, in his future career, as a 'mathematician in industry' in order that I should not be accused of avoiding difficult issues. Thus we must now decide what sort of preparation a mathematician in industry requires.

This preliminary inquiry aids us a great deal; indeed, we waste no effort at all in first coming to grips with the problem of the necessary equipment for the mathematician in industry, for I assert the proposition that to prepare a student to become X is a proper subset of the task of preparing a student to educate future X's. From this proposition follow two corollaries. The first, already mentioned, is that in this analysis we can leave on one side the task of the college mathematics teacher of educating future teachers since the solution to that question follows by induction once we have solved our given problem. Second, and more controversially, it follows that whatever the mathematician in industry requires will be required of the college mathematics teacher. (But more will be required of the latter.)

At this stage a little notation would perhaps be helpful. Let A represent the set of mathematicians teaching in graduate school. Let B be the set of graduate students in graduate school who will go on to become college mathematics teachers. Then the members of B are the students of the members of A and are to be thought of as the future teachers in colleges. Let C be the set of students in colleges. These then are the future students of the members of B and in our discussion so far we have been considering the future careers of the members of C. The question before us is how should the members of A influence and develop the members of B. Our argument has been that the members of B must be so educated that they can play any of the future roles of members of C, and that the predominant such role is that of a user of mathematics; and we will further argue, of course, that their training must include further components. By this notation we emphasize that the ultimate responsibility rests with the members of A—and this I believe to be one of the most important points to be made.

Let us return then to the question, what is the necessary education for the future mathematican in industry? Here I entirely concur with Herstein that it must consist of a thorough grounding in the art and science of mathematics together with a proper conception of, and real experience of, applying mathematics. It does not appear to me to be of great importance to what field of science the mathematics is applied, so long as the applications are genuine and significant. Thus I do not have in mind at all a traditional applied mathematics course. It is essential that the student himself should have experience of the various stages of work in applied mathematics. I list the stages under six headings:

1. The selection of a problem suitable for attack by the methods of mathematics.

2. Selection of a suitable mathematical model.

3. Collection of data.

4. Reasoning within the model.

5. Calculations.

6. Reference back to the original (non-mathematical) situation to test the validity of the conclusions drawn from the model.

Of course one is not to think of these six stages as simply forming part of a linear progression; there is considerable feedback in the process. For example, if it is discovered that the proposed solution is not valid for the original problem, then it may very well be necessary to modify the mathematical model or to collect more data. It may also be the case that the model is found to be too complex to enable any effective reasoning to take place. Then the model may be simplified (e.g., by linearization) in the hope that the new model will be susceptible of mathematical analysis and will still not be too far removed from the original problem.

I would like to remark at this point that a schema outlined above is valid, with certain obvious modifications, for research in what is known today as 'pure mathematics'. I would be very happy to see the distinction between pure and applied mathematics reduced to appropriate and moderate proportions. For me the distinction rests very largely on the issue of the type of problem which stimulates and motivates the mathematician, and not on the sort of mathematics he uses or creates. I believe that the procedures adopted by pure and applied mathematicians are, in principle, extremely similar.

Thus, it is necessary to emphasize that the fact that a student will expect to be applying his mathematics to the real world does not imply that he can or should ignore certain parts of mathematics. Of course, it implies that he should have a working knowledge of probability and statistics and an understanding of the role of computers. However, it would not be safe nor sound educational policy to cut him off from contact with algebra, topology and geometry simply because he is to become, in some sense, an applied mathematician. Examples abound today—Herstein gave some in his presentation—of parts of mathematics developed for purely analytical purposes which have turned out to be the appropriate tools for the study of problems drawn from the physical, biological and social sciences.

My remarks also imply that I am opposed to a trend clearly discernible today of appointing applied mathematicians to mathematics departments in colleges and universities in preference to excellent pure mathematicians, simply on the grounds that they will be expected to teach applied courses. I believe that any well-educated mathematician should be able and willing to teach service courses at the undergraduate level in any area of mathematics, and even any undergraduate courses to mathematics majors at least up to junior level. A mathematics department should try to get the best mathematicians available, whether they are 'pure' or 'applied'. It is of course perfectly legitimate, and indeed correct, to explain to a candidate for a position that he will be expected to teach, shall we say, statistics courses to psychologists. If he does not want to do this, then there is a very good case for moving on to the next candidate; but, to do as many departments are doing and to eliminate pure mathematicians at the outset because of the need to provide such service courses is, in my view, entirely wrong-headed.*

Who will be giving the graduate student the experiences and attitudes which we have said he should have in order that he should have the potential to become an effective college mathematics teacher? Returning to our earlier terminology, these will be the members of A. Thus I believe that a very heavy responsibility rests on the mathematicians in graduate schools today. It is simply not good enough to continue to do what we have done in the past, namely, to reproduce our own kind. We have to set an example to our students which will indicate to them clearly that we rate highly the tasks which they are likely to have to carry out for themselves in their own careers. We must show our respect and concern for applied mathematics. We must learn something ourselves of how mathematics is applied. We must also, of course, show our respect and concern for good teaching.

This, then, brings us to the question of what are the extra requirements that we should ask of the college mathematics teacher over and above what we ask of his successful students. That is to say, what should we, as members of the set A, be doing for our students as members of the set B, beyond preparing them to become, were they so minded, users of mathematics. These extra requirements arise from the fact that the college teacher will be teaching at various levels, so that he must be familiar with those levels and with the difficulties and natural thought processes of students at those levels. Of course it is common ground that any teacher of mathematics must be familiar with the topic he is teaching at a level above that at which he is teaching it. This alone would justify his studying some area of mathematics at the research level. Further, we must expect the future college mathematics teacher to have a real concern for teaching, and an understanding of what is basic to mathematical education. He can evince and develop his concern for teaching during his work in graduate school, and he must be encouraged to do so—this is the point that we were making above when we referred to the example which should be set

---

*I have not discussed here the perfectly legitimate desire a department may have to build up graduate work and research in some area of applied mathematics. In that case, of course, the department is perfectly justified in appointing the best person available in that particular field of mathematics.

by professors in graduate schools. He should also, in my view, acquaint himself with the ideas of those who have thought carefully about problems of mathematical education, and whose views command respect. I would go further and say that it would clearly be desirable that *all* graduate students should show an interest in questions of teaching. If for no other reason, it is surely the case that the best way to discover whether one understands some part of mathematics is to be placed in the position of having to teach it.

It would appear that I am saying that the experiences which should be provided in graduate school should include all those provided now and several more as well. I do not believe that one can shirk this conclusion. However, it does not frighten me too much. For it is an unfortunate fact that many graduate students spend an unconscionable time in graduate school, and that the time is very often not well spent. I believe it is necessary for us, members of A, to take far more responsibility and far more consistent responsibility for the education of our students. I have known all too many cases in which the nominal research supervisor is unable to state on what problem his nominal research student may be working. This neglect of the student by his supervisor could never be justified and today is less defensible than ever. We should make sure that every student in graduate school deserves to be there, and that he is using his time effectively. If this were the case, then I believe that it would be perfectly possible to fit into his graduate training those extra requirements which are indicated by my remarks above.

Let us admit that we have not been turning out good teachers at the pre-graduate level—or, at any rate, let us admit that if we have turned out such good teachers, it has been largely by accident. We have been turning out too many narrow specialists, people who are research-oriented without being necessarily research-talented. We have to provide experiences in the graduate school which will develop the many facets of mathematical talent, and which will lead our students to face the prospect of college teaching with enthusiasm, even in situations in which there will be little overt encouragement to them to do research. One way I would propose in which we can ourselves provide this encouragement is by taking the view that our students remain our special concern even after they obtain their doctorates and take their first job. We should not take the comfortable view that we are doing our duty provided we update from time to time our letters of recommendation for them. We should continue to concern ourselves with their development as mathematicians. We should seek to maintain personal contact with them, and we should seek to involve our own institution in their continued activity as mathematicians.

But here I go beyond my brief. I hope that my remarks will stimulate discussion of what is, by common consent, a very urgent and very important problem.

PANEL: THE ROLE OF APPLICATIONS IN THE TEACHING OF UNDERGRADUATE MATHEMATICS

Moderator: S. K. Stein

A BELATED REFORMATION

S. K. Stein

For centuries mathematicians have occupied the pinnacle of prestige. We were described in 1734 with these words in Bishop Berkeley's *The Analyst:*

> "...you, who are presumed to be of all men the greatest masters of reason, to be most conversant about distinct ideas, and never to take things upon trust, but always clearly to see your way... It is supposed that you apprehend more distinctly, consider more closely, infer more justly, conclude more accurately than other men..."

Since World War II we have tacitly agreed. Swelling college enrollments and research support sustained our self-esteem. There was no compelling reason to take a hard look at ourselves. But in recent years the employment crisis has shattered the self confidence of the mathematical community.

In the brief golden era that stretched from around 1950 to 1970 the value of pure mathematics—largely detached from its origins and its applications—was seldom questioned. New and growing departments made sure that pure mathematics was well represented. Generally, neither undergraduate nor graduate mathematics students were required to study physics or any field that traditionally used mathematics. The new-math revolution reflected the triumphs of the pure mathematician: what was valuable at the frontiers of research was assumed appropriate to every child in the land. One high school curriculum project was based on the announced assumption that 50% of the populace would require calculus.

Now that the tide of research support and student enrollment has turned from pure to applied math, departments have moved in unison, from starboard to port, tilting the ship of mathematics precariously in the opposite direction. Computing and statistics split off from some math departments, or at least develop their own majors, leaving the traditional math offerings unchanged, and sometimes unattended. A speedy abstract algebra course might be offered in the electrical engineering department, while enrollment in the traditional one dwindles.

Some mathematicians are trying to return applications to the standard curriculum. They are doing this not just as a projection upon their students of their professional frustrations. After all, the broad undergraduate curriculum should not simply reflect the latest forecast for Ph.D. employment. Indeed, all through the "new-math" movement, which ranged from kindergarten through college, some mathematicians did criticize its alienation from the origins of mathematics. How do we bring back applications from their exile? Should we replace whole courses with new ones? Must we insist, not merely recommend, that undergraduate math majors take a sizable dose of physics and other disciplines that use mathematics, perhaps even a double major? Should we include applications in our traditionally "pure" courses, even at the sacrifice of some theory?

Is there a danger that the applications we devise or choose ourselves, on the basis of a summer's crash seminar and the conceit Berkeley described, might have primarily an esthetic appeal? Might we not create—out of pedagogical zeal—a new branch of knowledge, "pure applied mathematics," which is pure math camouflaged in the applied jargon? Such so-called applications are never put to the test of real applied math, namely, that the model tells us more than we told it. Is it reasonable to expect a generation of mathematicians, narrowly trained, perhaps oblivious of the origins of their own field of interest, to convey to their students a broad view of mathematics?

We should be wary of choosing as "applied math" some new exciting advance in pure mathematics, which claims eventual applications. The rise and decline in the theory of games should stand as sufficient warning. Nowakowska ([2] p. 267) after a detailed study of publications in this field, concluded: "The second factor which influenced the development and the loss of interest in theory of games is connected with the change of interest in [the] theory of games by non-mathematicians. Since the starting point of the theory of games is the analysis of situation [s] of conflict, it was generally expected that this theory would be applicable to [the] study of such situations in social sciences. This caused originally a great demand for mathematicians dealing with theory of games: [who] were probably financed by army, and other institutions interested in potential applications of theory of games to their problems. Thus, the interest in [the] theory of games, at least in the United States, was reflected in financing mathematicians. As it gradually turned out that theory of games cannot serve as satisfactory universal model of situations of conflict, the interest of non-mathematicians (and consequently, also that of mathematicians) faded." Contemporary candidates for a similar evolution may be the theory of fuzzy sets and catastrophe theory.

I think we need not wait for the next generation of mathematicians to begin to compensate for our failures of omission. Before making some modest suggestions—more precisely, raising what I think are questions worth answering, let me cite some typical attempts already made in this direction.

One professor, teaching an upper-division algebra course, thought it would be a good idea to devote the third quarter to the many applications of algebra in coding theory. His department turned down his proposal, insisting on the traditional application: Galois theory.

As a calculus author, Morris Kline incorporated physical applications in such a way that they could not be skipped. Years later he observed in a letter, "most college professors do not wish to teach physical problems. Many are not prepared to do so and others just do not wish to give it the time. They think that they make more progress by teaching more calculus, even if the stuff is meaningless to the students."

Another calculus author put the applications in separate chapters. Professors did not cover them. In the next edition, applications infiltrated the book in separate one-day sections. Still they were generally skipped. Users advised the publisher to "keep them in, but we don't have time to cover them." Gradually the applications were demoted to "examples," thence to "exercises," and lastly to the final resting place of "review exercises."

There are now at least two upper-division algebra texts that incorporate significant applications, and at least one differential equations text that motivates each chapter by application and history.

For twenty years one college offered a course, taught jointly by physicists and mathematicians, which integrated calculus with physics. One professor, looking back on the experiment, commented, "Mathematics suffered from this arrangement. Students tended to think of mathematics as existing solely in the service of physics, which was perceived as the more modern and exciting of the subjects."

One department introduced a course on applications and a course on history, but few can teach the first and no one knows what to do in the second. Both are fundamentally guilt offerings to compensate for defects in the "standard" courses.

These few cases show that some mathematicians have been trying to wed theory and application after their long separation in the curriculum. But they also warn us that the effort to do so may fail or even boomerang. Still, there are some modest steps that any mathematician, as teacher or author, might consider, steps that might be dubbed an "affirmative action program" to put mathematics in proper perspective. My concern is for the short range period. The experiences we have with a variety of small experiments will presumably suggest more substantial proposals in the years to come.

First of all, whether conducting a service course or a course for math majors, the instructor can ask, *"How is this material related to other branches of mathematics or the sciences? Could I spare a few days of the course to introduce these applications?"* These questions may raise difficult choices: Should I cut down on homomorphisms in order to include the Burnside counting theorem? Should I sacrifice related rates or curvature in order to cover Poisson traffic? Should I omit the proof of a theorem on finite fields in order to explore their many combinatorial and geometric applications?" Even if these choices are not articulated they are made implicitly every time we teach. By teaching again as we taught before, we are saying, "emphasize the theory; applications will take care of themselves, somehow, later."

Answering the questions may require a little background reading on our part, perhaps preparing a ditto or a xerox from some text, or putting a reference on reserve in the library. But even the busiest instructor can find the time to do at least a little in this direction. We need not wait for textbooks to make the changes for us. The *Monthly,* for instance, in the last few years has presented many attractive topics.

A student who takes even a pre-college algebra course with us should know—during the course—the answer to the question, "Why would anyone ever have to solve a quadratic equation," even though he may have been trained not to ask it. A biology or economics student who takes a calculus course should see at least one convincing example that relates calculus to his major. Professors in these fields have remarked that even one example can go a long way in sustaining such a student throughout the course. An upper-division algebra student should be able to answer the simple question, "What good are groups?" We should assume that even our own captive majors will be handicapped in their studies and teaching by our disregard of origins, links, and applications. A textbook author, like an instructor, is a public-relations representative of the mathematical community. He—or she—faces analogous questions: *"Will a student reading my book get a fair idea of the significance of the mathematics I present? Have I done all I can to relate theory and practice? Must a student go on to the next course in order to appreciate the course that my book serves?"*

The mathematical community, threatened in a way that would astonish Bishop Berkeley, is trying to cultivate a better public image. It even assigned one of its members, for one year, to get our message into the media. But the classroom and the text, through which we reach millions of students day after day, challenge each of us to convey a balanced picture of mathematics, showing its intrinsic beauty and vitality but also its diverse and frequently unexpected applications in the mathematical realm and in the real world. No major curriculum overhaul may be needed to accomplish this task. But we will, as individuals, have to admit our bias, nurtured in our training, and change our habits. May the reformation last long after the present job crisis passes.

But when we make changes in our teaching, let us not lose sight of a crucial function of mathematics: its value in developing clear and logical thinking. Perhaps the following words may remind our students and ourselves of this quality in our discipline. Conrad Hilton in ([1], pp. 63–64) observed:

> "I'm not out to convince anyone that calculus, or even algebra and geometry, are necessities in the hotel business. But I will argue long and

loud that they are not useless ornaments pinned onto an average man's education. For me, at any rate, the ability to formulate quickly, to resolve any problem into its simplest, clearest form, has been exceedingly useful. It is true that you do not use algebraic formulae but in those three small brick buildings at [my college] Socorro I found higher mathematics the best possible exercise for developing the mental muscles necessary to this process.

In later years I was to be faced with large financial problems, enormous business deals with as many ramifications as an octopus has arms, where bankers, lawyers, consultants, all threw in their particular bit of information. It is always necessary to listen carefully to the powwow, but in the end someone has to put them all together, see the actual problem for what it is, and make a decision—come up with an answer. A thorough training in the mental disciplines of mathematics precludes any tendency to be fuzzy, to be misled by red herrings, and I can only believe that my two years at the School of Mines helped me to see quickly what the actual problem was—and where the problem is, the answer is. Any time you have two times two and know it, you are bound to have four."

Confident but not dogmatic, we will find that applications, judiciously chosen, will enrich our classes and broaden our own perspectives.

References

1. Conrad Hilton, Be My Guest, Hilton Hotels, pp. 63-64.
2. M. Nowakowska, Epidemical spread of scientific objects: an attempt of empirical approach to some problems of meta-science, Theory and Decision, 3 (1973) 262-297.

APPLICATIONS TO THE PHYSICAL SCIENCES

A. B. Willcox

Have you heard the one about how calculus helps you cough? No? I'll tell it to you.

Picture your lungs as a balloon inside your rib cage; a balloon within a balloon so to speak. Your bronchi and trachea are like a tube connecting the inner balloon, your lungs, with the outside world. When you cough your rib cage contracts rapidly, sharply increasing the pressure inside. This pressure, creating an equal pressure inside the inner balloon, causes air to be expelled rapidly through bronchi and trachea to the outside—in a manner occasionally disturbing to your immediate neighbor. This is a cough. Now, a certain part of the tube connecting lungs to the outside is elastic. Let us call this elastic portion the bronchi. That same pressure which is expelling the air from your lungs is tending to collapse the soft bronchial tube, thus constricting the outward flow of air. Nature therefore seems to be working against itself. The pressure within your rib cage simultaneously drives air out of your lungs and chokes off the outward flow. How is it, then, that you are able to work up a good cough? How inefficient of nature.

Perhaps a mathematical model of the mechanism of a cough will help. Let the coughing pressure within the lungs be P (actually, the pressure differential, but let us assume 0 pressure in the outside atmosphere) and let the radius of your bronchi be r. Let the normal, or rest, radius of the bronchi be r_0 and let v be the velocity of the air flow through bronchi during the cough. In order to construct a mathematical model we must make some assumptions about the nature of reality. If the model is to be useful they had better be reasonable assumptions. In this case it seems reasonable to assume that the bronchi are elastic, within a range of distortion, and that the air flow through the bronchi is that of a perfect fluid through a cylindrical tube. Under these assumptions, physicists tell us that the distortion in the radius from rest, that is $r_0 - r$, is proportional to the pressure P and that the velocity of flow is proportional to both P and the square of the radius. From the resulting two equations in P, r, and v, we may eliminate P, ending up with an equation of the form

$$v = k(r_0 r^2 - r^3),$$

where k is a combination of the two proportionality constants.

Now ask your calculus students the conditions under which the velocity of air flow is maximized. Even the slow students will tell you that v is maximum when $r = 2r_0/3$. X-ray observation confirms that during a cough your bronchi are indeed collapsed by about a third in radius. Thus, we learn from this model that nature is not at all inefficient. The purpose of a cough, after all, is to create a high velocity air flow through bronchi and trachea in order to dislodge and expel some foreign body. Despite your bumbling nature you seem to be able to do the best possible job of it. And without calculus you would go through life ignorant of your accomplishments!

Now that I have your undivided attention, let me acknowledge and attempt to engage my assigned topic in this panel discussion. We will return to COUGHING WITH CALCULUS shortly. Actually, without admitting it to the organizers of the panel, I have planned from the beginning to take great liberties with my topic. I will make a try for legitimacy by returning to the special role of the physical sciences, but my strongest opinions about the role of applications in the teaching of mathematics to undergraduates are in no sense restricted to one field of application.

I can perhaps best illustrate what is on my mind by describing how one of my favorite quotable quotes has fallen from grace. I have long since forgotten the source, but about 20 years ago I was first amused, then pleased, then more profoundly influenced by a statement attributed to a contemporary: "There is no applied mathematics, only applied mathematicians." Trite, perhaps, but I adopted it as a sort of personal creed. I still accept it, but lately it has lost some of its luster. It needs to be updated. All creeds need periodic rejuvenation because as individuals we accept them according to the private meanings we read into them. Times change, and with them our perceptions. A new message has lately surfaced in this quotable quote that may always have been there but was submerged by my sympathetic vibrations. It seems to suggest that the mathematical community is divided into disjoint sets; pure mathematicians and applied mathematicians. Well, it just isn't so, and I can live no longer with the tarnished creed. It must be changed. I am not very good at producing aphorisms, but if I had to distill my present belief into a few words I would come up with something like this: "There is no applied mathematics, there are no applied mathematicians, there are no pure mathematicians, only mathematics and mathematicians."

Some of us *do* mathematics by choice, others prefer to *use* mathematics, but most mathematicians have mixed tastes. Show me the "applied" mathematician who does not often delight in the pristine beauty of the abstract mathematics he uses to advance his understanding of problems in the real world. Show me the "pure" mathematician who isn't at least occasionally fascinated to see the patterns and relationships he deals with in abstraction arising in the real world. Yes, I know, you *can* produce examples of the pure and the purely impure; every spectrum has its extremes. But mathematicians spread all across the rainbow and, I daresay, in something like a normal distribution. And with mathematicians' tastes, so goes mathematics.

Our discipline, like all disciplines, is an island, in a sense self-contained. But it is *not* isolated. Bridges join our island to other islands in the sea. Many fields depend heavily on mathematics. Mathematics is itself absolutely dependent on the bridges joining it to these other fields. Traffic on these bridges is two-way, carrying indispensable tools for organizing ideas and reaching conclusions in one direction and a disorganized mass of observed data in the other. The tools help society; the mass of data contains hidden patterns that stimulate and direct the creative imagination of the mathematician. Mathematics thus lives in a healthy symbiosis with the other disciplines.

If our colleges and universities are to train students in living mathematics, not the encrusted tracks we mathematicians leave as we wander about our island, then this symbiosis must enter the classroom. Students should be led to explore not only the island mathematics, but also the bridges linking it to other islands.

This brings me at last to my assigned topic. These bridges are of course applications of mathematics to other disciplines. When applications can be found at the appropriate level and involving the appropriate mathematics, they should be explored in the classroom with enthusiasm on the same intellectual level as the mathematics itself. I am convinced that if we do not recognize this imperative we will raise a new generation of provincial islanders and start mathematics on the lonely road to isolation and eventual decay.

The bridges we use in the classroom are, of course, miniatures of the great bridges that supply tools and ideas for most of the intellectual world and bring stimulation and direction to mathematics. How does one select these classroom bridges? I will offer several guiding principles that may be helpful to undergraduate teachers. They constitute the most concise description of an effective classroom application that I have been able to devise.

1. *A classroom application must be interesting to the teacher and to the student*. This is a subjective criterion not independent of the other guidelines I will list. It is also a perfectly obvious one but it is an essential guideline. Your students will not learn if you put them to sleep. I mention the teacher first because the excitement must begin with the instructor. What

you find dull and unimportant usually, if not always, is perceived by your students to be dull and unimportant. You must have faith that the reverse is also true; what you find fascinating will excite your students. There are many useful applications that are dull reading by any standards. You can recognize them. Avoid them. The most effective applications contain an element of surprise or a moment of enlightenment. The object of application of mathematics is discovery of something new about one of the islands it touches.

2. *A classroom application need not itself make an immediately useful contribution to another field, but it should clearly illustrate the potential for transmitting significant ideas to that field or to mathematics.* The specific problem and the particular mathematical model may be insignificant and oversimplified, but they ought, at the same time, to suggest and illustrate a useful class of serious application. Your students are intelligent enough to extrapolate beyond the walls of your classroom. You should be careful not to claim too much for the specific example you use, but don't be overly modest either in describing the potential of related applications.

The bridge to the outskirts of physiology which I described in my opening remarks is a case in point. I hope you will agree that it satisfies the first guideline. It should capture the interest of the students. At least I find it interesting, partly because it is unusual and partly because it can be presented so as to provide a mild surprise at the end. Nature is efficient after all! The application is far too simplified to be really serious, perhaps. One can hardly imagine a physiologist exclaiming "Eureka" as he sets the derivative equal to 0 and examines the solution. But it does illustrate a medium for using mathematics to help scientists understand the mechanics of the human body, and it may therefore sensitize the student to some emerging bridges to serious mathematical biology.

3. *A classroom bridge should touch some part of a student's home territory.* A bridge is of no use to a person who cannot reach it. The application must pertain to a problem that the student can understand and will recognize as at least marginally part of his or her world of experience. This may sound a bit restrictive, given the diversity of many undergraduate classes. If it is not given a liberal interpretation it may exclude many of your favorite applications. But don't underestimate the breadth of your students' worlds. They extend almost to the limits of their imaginations which, you will agree, are precocious. Most students know a little about their own bodies, for example, most know small bits of elementary physics, and all have coughed.

This guideline gives me the opportunity to complete my assignment to comment on the special role of applications to the physical sciences. There are many rapidly emerging areas of applications of mathematics in a wide range of fields—social, political, psychological, economic, and many others—bearing on problems of great contemporary societal importance. These should be represented in the classroom if for no other reason than that they command instant attention from the students. However, physical ideas such as velocity, acceleration, energy, or momentum, are part of the universal experience of living, and therefore applications to the physical sciences will probably always be preeminent in the teaching of most mathematical subjects, particularly those in the analysis core. It was my good fortune to participate in an interdisciplinary calculus-physics course taught to freshmen at Amherst College for 15 years in the 1950s and early 60s. That was a long time ago and much water has gone over the dam since that course was swept away in a flood of curriculum revision in about 1965. But the excitement of that rich introduction to the interplay between mathematics and physics haunts me to this day. Anyone who has traveled in an automobile or plane, watched a rocket rise majestically from its pad, seen earth satellites of more than the green cheese variety, witnessed a space walk, seen the moon close up or the earth from far away, can appreciate the ideas of velocity, acceleration, momentum, kinetic energy; can appreciate Newton's laws of motion; can see Kepler's laws of planetary motion at work in our own back yard; can understand escape velocity. The ideas have been constantly in the newspapers for two decades. Within the limits of time, all of these ideas belong in a calculus course, both to illustrate the power of mathematics and also to motivate and develop the mathematical ideas themselves. The second purpose is as important as the first, and the observation brings me to my fourth and final guideline.

4. *A classroom bridge can be equally useful traveled either way.* This is why I wish we had a better word than "application." "Application" suggests the use of the power of mathematics to solve problems and organize ideas in other fields. This process is of the greatest importance to mankind, of course, but equally important—for mankind as well as for our little island—is the flow of ideas in the other direction. That is why I keep referring to "bridges." Throughout history mathematics has been stimulated and, to a large extent, directed by its contact with other fields. Without this continual commerce with other islands, mathematics, like any single discipline, will degenerate into a complex game, with rules becoming too complicated and demanding for any but the professional player and too technical to attract even the casual spectator. Mathematical ideas do not spring independently into the minds of mathematicians. They are extracted, or abstracted, from existing patterns and ideas, often originating outside of mathematics. Ideas should enter the classroom the same way, by a process of abstraction from something meaningful to the students. In other words, every mathematical idea should be approached along a bridge from wherever the student is. A large number of applications you use in your classroom—perhaps half of them—should be selected for the light they shed on mathematics.

The importance of this guideline was driven home to me recently by the following experience. A good friend wrote in a state of high dudgeon about some curricular materials we had been discussing related to the mean value theorem. The object of his scorn was an application of the mean value theorem to prove that a motorist who traveled between two particular points in a particular length of time must certainly have exceeded the speed limit at some time during the trip. "What a travesty," growled my friend—I paraphrase his remarks. "For reputable curricular materials to suggest that a person reach for the mean value theorem to argue such a point is an intellectual fraud. There are many more natural and more intuitive ways to argue that if an object's velocity is smaller than its average velocity some of the time during a trip then its velocity must be greater than the average at some other time. The mean value theorem was invented for far nobler purposes. Such nonsense will portray elegant and powerful mathematics as a paper tiger."

His point is well taken, and I partly agree with him. But in another sense I completely disagree. If the "application," which I would prefer to call a "real world connection," is presented properly, it has a useful place in any treatment of the mean value theorem. What does the theorem state, after all? It states that the average velocity of an object must be attained at some point during the trip. At some time, the instantaneous velocity must equal the average velocity. What could be more simple, more easily remembered. The analytical statement of the mean value theorem is a mess of technical jargon to the average student until he has lived with it for a while, but the speeding motorist will not quickly be forgotten. The statement about velocities to which this episode leads is natural and intuitive, and it tells the story of the mean value theorem at least as accurately as the familiar one about chords and tangents. The student who has discovered the theorem in such a familiar context will consider it a friend and will be well on the way to understanding. Let the speeding motorist *introduce* the mean value theorem in your classroom, don't ask him to *justify* it.

I might say, in further recognition of my assigned topic, that because of the traditional close ties between mathematics and the physical sciences, you will probably find a majority of such "motivating bridges" leading in this direction. We certainly found in the Amherst interdisciplinary course that with minor adjustments in the arrangement of the calculus material we always had significant problems from physics ready at hand to introduce each new calculus concept as it came along. However, I expect to see this center of gravity moving in the next decade toward some of the newer areas of application. There are already some exciting examples from these areas, as our next speaker will undoubtedly point out.

In my remarks today I have almost totally ignored most of the hard questions concerning the introduction of more applications into the undergraduate curriculum. Where does one find them? Where does one find the time for them in the busy academic calendar? Does one give up some actual mathematics in order to present more applications? What role should our academic colleagues in other disciplines play in the selection and presentation of applications of mathematics? I've neglected these questions partly because of the nagging of the clock and partly because of the limitations on my own competence. I am hardly competent to answer some of these hard questions of implementation, and I am sure that if I were transplanted into the classroom today I would be as perplexed by them as any of you. But my ignorance does not indicate that answers cannot be found showing the way to a healthy mix of theory and application in our undergraduate curriculum. I think they will be found, *you* will find them because they *must* be found.

DNA COUNTS, PESTICIDE PROJECTIONS, AND VEHICULAR VECTORS: APPLICATIONS OF UNDERGRADUATE MATHEMATICS

Fred S. Roberts

1. Introduction. Imagine the mathematician of 1976, dressed in a lab coat, counting DNA molecules. In a swamp off campus, a colleague is projecting levels of pesticides. The chairman of the math department is standing at the corner of Fifth and Main, vectoring vehicles through the rush hour traffic.

Is it possible that these people are doing mathematics? Of course not. The dress and the setting are not appropriate. But mathematicians are increasingly interested in applying the tools of their trade toward the solution of problems of society. Among those societal problems which mathematicians (and non-mathematicians using mathematical tools) are attacking are problems involving energy, transportation, pollution, and health care delivery. I would like to argue, by considering the mathematics of DNA counts, pesticide projections, and vehicular vectors, that the expanding horizons of mathematical applications have significant implications for undergraduate mathematics.

The mathematics which is used most frequently in societal applications is strikingly simple. There are several reasons for this. First, the imprecise nature of societal problems: societal problems often involve hard-to-quantify, imprecise relationships, and the simplest types of models are the most appropriate for describing these relationships. Second, the complexities of societal problems: these problems tend to involve many variables, and quickly lead to extremely difficult mathematical problems unless major simplifying assumptions are made. Thus, *by necessity*, mathematical treatments are vastly oversimplified.

Of course, it is incumbent upon modern mathematics to develop tools and new types of mathematical techniques which are relevant to these problems of society. However, that is not my main point. My main point is that these observations have implications for undergraduate mathematics education.

2. Implications for undergraduate mathematics. Three of the implications for undergraduate education are the following. First, mathematical techniques, even the most elementary ones, can cast light on a real-world phenomenon. Second, problems with no "nice" solutions *should* be brought into the classroom. Third, mathematical problems, simple to state but at the frontier of modern research, *can* be brought into the classroom. I will illustrate each of these points with an example. All three examples are relevant to an elementary finite math course. They could obviously be used in a probability course, in a modelling course, etc.

3. DNA counts. Mathematical techniques, even the most elementary ones, can cast light on a real-world phenomenon. To illustrate this first point, let us consider the phenomenon of the vast complexity of nature. This is not exactly a societal problem, but it is related to some important ones such as social decisions involving eugenics, and environmental decisions involving the impact of policies which could cut down on diversity in an ecosystem.

The simple explanation of the complexity phenomenon, on an intuitive level, involves nothing more than the multiplication rule: if something can happen in n_1 ways, and for each of these a second can happen in n_2 ways, and for each of these $\ldots$, then the sequence can happen in $n_1 \times n_2 \times \ldots$ ways.

Genetic information is encoded in the DNA molecule. DNA is a chain of chemicals (called nucleotides), each of which is made up of a phosphate, a sugar, and one of four bases. The important genetic information is contained in the sequence of bases. Thus, DNA can be thought of as a chain each link of which is one of the four bases Thymine (T), Cytosine (C), Adenine (A), and Guanine (G). A typical (portion of a) DNA chain could be represented as ATTAGGCGCTA.

We begin by asking how many 2-element DNA chains there are. The answer is $4 \times 4 = 4^2 = 16$, by the multiplication rule. How many 3-element chains are there? There are of course 4^3 or 64. How many chains of 100 elements? There are 4^{100}.

To illustrate the power of just 4 bases to encode genetic information, let us ask what a typical DNA chain is in length. In a chicken, there are 5×10^{19} bases per DNA chain. That means that there are $4^{5 \times 10^{19}}$ possible DNA chains. Now it is simple to estimate that $4^{5 \times 10^{19}} > 10^{3 \times 10^9}$, and so we see that there are more possible DNA chains in a chicken than one followed by 3×10^9 zeroes! Similar information for other organisms is contained in Table 1. The reader will quickly see the amazing number of possible DNA chains. It is not surprising, then, to see such variety in nature.* The simplest mathematical technique has illuminated this point.

---

*More detailed discussion of this example and the next can be found in Mosimann [1]. The author thanks Professor Helen Marcus-Roberts for introducing him to these two examples.

Table 1. Lengths of DNA Chains

Organism	# bases per DNA chain	# possible chains
chicken	5×10^{19}	$4^{5 \times 10^{19}} > 10^{3 \times 10^{9}}$
mouse	1.3×10^{10}	$4^{1.3 \times 10^{10}} > 10^{7.8 \times 10^{9}}$
guinea pig	1.7×10^{10}	$4^{1.7 \times 10^{10}} > 10^{1.02 \times 10^{10}}$
human	2.1×10^{10}	$4^{2.1 \times 10^{10}} > 10^{1.26 \times 10^{10}}$

4. Pesticide projections. The second implication for undergraduate mathematics education which I would like to illustrate with an example is that problems with no "nice" solutions *should* be brought into the classroom.

Many "modern" uses of mathematics involve aids in making decisions. Mathematical tools are used to build models which forecast or project the effects of certain activities, the impacts of certain policies, etc. A typical place where such aids can be useful is in decisionmaking about pesticides or other chemicals which enter the environment. Pesticides such as DDT are especially dangerous because of their longevity: they stay in the ecosystem for a long time after being introduced. How could we project the relative longevity of two different pesticides?

There are obviously many factors relevant to the flow of a molecule through an ecosystem. Probably the most useful model of this flow would involve some understanding of why and how the molecule moves from one part of an ecosystem to another. But the mechanism behind this movement is not well understood.

I would like to argue that one can gain some insight into the relative longevity of different substances in ecosystems by making rather significant simplifying assumptions. I shall build a very simple model. In the model, an ecosystem will be thought of as having various states. Following Mosimann [1], let us deal with a pasture ecosystem, and assume that there are four states: soil, grass, cattle, and "outside." Once a DDT molecule* is in one of these states, we assume it can do one of several things: stay there, or pass on to another of the states. The possibilities are shown in Figure 1. We see that once the molecule is in the soil, for example, it can either stay in the soil, be absorbed by the grass, or erode out of the ecosystem. Once a molecule of DDT is outside the ecosystem, we assume that it stays outside from then on.

We look in on the ecosystem every time period, say every minute, week, or month, or so. Rather than study the mechanism of flow of the molecule through the system, we simply ask what is the probability that a molecule which was in state x_i at one time period will be in state x_j the next time

---

*Mosimann traces a phosphorus molecule, but the principles are the same.

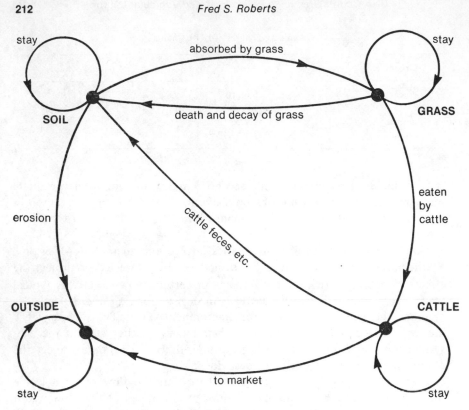

Fig. 1. Possible transitions for a DDT molecule in a pasture ecosystem.

period? (This is of course a conditional probability.) We denote this probability p_{ij}. Typical probabilities for our example are shown in Table 2, and in matrix form in Table 3.

Table 2. Probabilities for Pasture Ecosystem

$p_{SS} = .6$	$p_{SG} = .3$	$p_{SC} = 0$	$p_{SO} = .1$
$p_{GS} = .1$	$p_{GG} = .4$	$p_{GC} = .5$	$p_{GO} = 0$
$p_{CS} = .75$	$p_{CG} = 0$	$p_{CC} = .2$	$p_{CO} = .05$
$p_{OS} = 0$	$p_{OG} = 0$	$p_{OC} = 0$	$p_{OO} = 1$

Table 3. Probabilities for Pasture Ecosystem in Matrix Form

$$
\text{From}
\begin{array}{c}
 \\ S \\ G \\ C \\ O
\end{array}
\begin{array}{cccc}
 & \text{To} & & \\
S & G & C & O \\
\left(\begin{array}{cccc}
.6 & .3 & 0 & .1 \\
.1 & .4 & .5 & 0 \\
.75 & 0 & .2 & .05 \\
0 & 0 & 0 & 1
\end{array} \right)
\end{array}
$$

Using the conditional probabilities p_{ij}, we can calculate various other important probabilities. For example, given that the DDT molecule starts in the soil, we can calculate the probability that it will be outside the eco- system after two time periods. This calculation is made using tree diagrams, as shown in Figure 2.

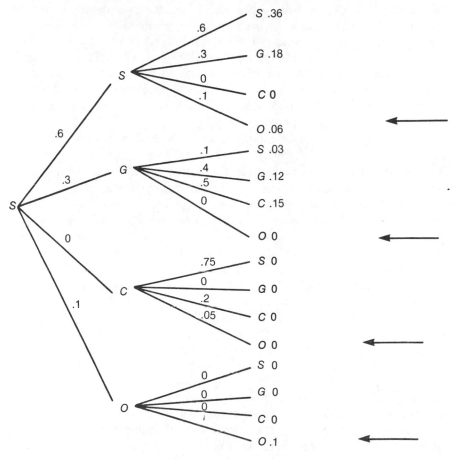

Probability out = .06 + 0 + 0 + .1 = .16.

Fig. 2. Calculation of the probability that a DDT molecule will be outside of the pasture ecosystem after two time periods, given that it starts in the soil.

A similar computation can be made for pesticides other than DDT, and the various probabilities can be compared. The information can be used as an aid in decisionmaking.

Somewhat more complicated questions can be answered if we view this model as a Markov chain. Then the state "outside" is an absorbing state. We can calculate by well-known techniques such values as the expected number of time periods before the DDT molecule leaves the system, given that it starts in the soil. (The process essentially just involves matrix inversion.) The result is more sophisticated information about the ecosystem. (For details of the computation, and more discussion of this example, see Roberts [3].)

All of this can be done in a finite math course. However, it should not be left at that, because students will certainly see the oversimplifications in the model. One should now discuss these oversimplifications. For example, we are assuming that there are well-defined "states" in the ecosystem. We are assuming that the conditional probabilities are time-independent and independent of the routing the molecule took to get to a particular state (assumptions we used implicitly). One should also look at the difficult practical question: how would one get the information needed to use the model? That is, how would one get the p_{ij}?

The answer to the simplifications is that one has to make simplifying assumptions since we don't understand the mechanisms involved. Moreover, bringing in other factors than those considered would make for a rather complicated problem. We make simplifying assumptions at an early stage of trying to understand a phenomenon. These assumptions have to seem "reasonable," and ours do. Even with such assumptions, the use of the model gives us some insight into the passage of pesticides through ecosystems. Students gain mathematical insight into the relationship between assumptions and conclusions. We all learn about how complex the operation of ecosystems is, and we pinpoint some of the questions which have to be asked about ecosystems in the future.

Answers to the practical question of availability of data are often guessed at by students. For example, some form of radioactive tracing might work. However, it is a good idea to point out in general that a model is only as good as one's ability to gather the data the model uses.

In any case, the main point is this: students should be allowed (indeed, encouraged) to think about the difficulties, and about how they might go about attacking them. In the process, they will develop mathematical ways of thinking, which will make them more careful, more precise, and hopefully better decisionmakers in the future.

5. Vehicular vectors. The third and final implication for undergraduate mathematics education which I would like to illustrate is that mathematical problems, simple to state but at the frontier of modern research, *can* be brought into the classroom. I would like to illustrate this by means of a one-way street problem. A town has a growing traffic congestion and air pollution problem. Someone has suggested that traffic would move better, and so smog level would improve, if all streets were made one way. But can this always be done? And if so, how?

To answer these questions, let us use graph theory. Assume that currently all streets in the town are two-way. Represent the locations in town as the vertices of an (undirected) graph, and draw an edge between two locations if and only if they are joined by a two-way street. An example of such a graph is shown in Figure 3. We would like to put an arrow or direction or orientation on each edge of this graph. Can this always be done? The answer is: of course. Simply put the arrows on at random. Unfortunately, this can lead to situations like that shown in Figure 4, which would be a fine orientation—for someone owning a parking lot at *a*! Vehicles must be vectored more intelligently than this.

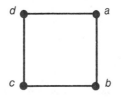

Fig. 3. A graph representing a two-way street network.

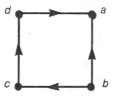

Fig. 4. A one-way street assignment for the graph of **Fig. 3.**

The problem needs to be defined more precisely. We would like to assign a direction to each street so that it is possible to get from any location to any other location. In the language of graph theory, we would like to orient each edge of the undirected graph so as to obtain a strongly connected directed graph as a result. Of course this can be done for the graph of Figure 3. A strongly connected one-way street assignment (orientation) is shown in Figure 5. However, such an assignment does not exist for every undirected graph. Figure 6 shows a graph which has no such assignment. Certainly the original graph must be connected. Figure 7 shows a connected graph with no strongly connected one-way street assignment. (Whatever direction is given the edge α leads to trouble. Either it is not possible to get back from *b* to *a*, or it is not possible to get back from *a* to *b*.) Figure 8 shows a similar example.

The edge α of Figures 7 and 8 has the property that its removal disconnects the original graph. Such an edge α in a connected graph is called a *bridge*. If a graph *G* has a strongly connected one-way street assignment, then clearly *G* must be connected and have no bridges. The following theorem was proved by Robbins [2]:

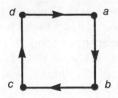

Fig. 5. A strongly connected one-way street assignment for the graph of **Fig. 3.**

Fig. 6. A graph with no strongly connected one-way street assignment.

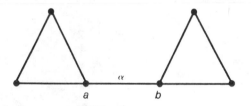

Fig. 7. A connected graph with no strongly connected one-way street assignment.

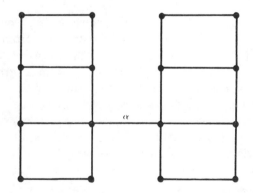

Fig. 8. Another connected graph with no strongly connected one-way street assignment.

Theorem. *A graph G has a strongly connected one-way street assignment if and only if G is connected and has no bridges.*

Robbins' Theorem answers the question of when a one-way street assignment with the desired property exists. But it does not tell how to find such an assignment. Along with this existence theorem, however, goes a simple procedure for finding one-way street assignments. The procedure is described in Roberts [3].

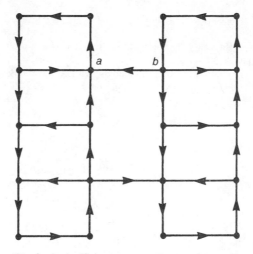

Fig. 9. An inefficient one-way street assignment.

There is a serious problem with the existing procedure for finding one-way street assignments. It leads to such street networks as that shown in Figure 9. A person living at location *a* who works at location *b* is very unhappy with this assignment! He has to drive a long distance. The assignment shown is inefficient in some sense. One wants a precise definition of efficient one-way street assignment, theorems about when such assignments exist, and procedures for finding them. Not much is known about these problems, and even for the simplest definitions of efficient, for example involving some notion of minimizing average or maximum distance travelled, the solution of these problems is not straightforward. Thus, a simple real-world question has quickly led to the frontiers of mathematical knowledge.

6. Conclusion. In conclusion, modern problems of society provide exciting opportunities to use mathematics. I think that a major goal of mathematical education is to train a large number of future decisionmakers to make intelligent decisions, both public ones and private ones. We can do this by bringing in real-world problems and realistic applications of mathematics into existing courses. Not all of these problems should have satisfactory solutions, nor should they all be neat, clean, and concise. Indeed, often more can be learned from problems without neat solutions. Even without perfectly satisfactory solutions, one can illustrate the power of mathematical thinking, and one can find an opportunity to introduce beautiful mathematical results, with a point.

References

1. J. Mosimann, Elementary Probability for the Biological Sciences, Appleton-Century-Crofts, New York, 1968.

2. H. E. Robbins, A theorem on graphs, with an application to a problem of traffic control, Amer. Math. Monthly, 46 (1939) 281–283.

3. F. S. Roberts, Discrete Mathematical Models, with Applications to Social, Biological, and Environmental Problems, Prentice-Hall, Englewood Cliffs, N.J., 1976.

INDEX